The Amateur's Workshop

Ian Bradley

*with chapters by
the late
Dr. N. Hallows
who also took
many of the photographs*

NEXUS SPECIAL INTERESTS

Nexus Special Interests Ltd
Nexus House, Azalea Drive
Swanley
Kent BR8 8HU
England

First published 1950
New edition 1971
Reprinted 1974
Revised edition 1976
Reprinted 1979
First paperback edition 1984
Reprinted 1986, 1987, 1988, 1995, 1999

© Argus Books 1976

ISBN 1-85486-130-1

All rights reserved. No part of this publication may be reproduced in any form, by print, photography, microfilm or any other means without written permission from the publisher.

Printed and bound in Great Britain by
Biddles Ltd, Guildford & King's Lynn

Contents

Chapter		Page
1	The Workshop	5
2	The Lathe	19
3	The Drilling Machine	32
4	Belt Drives	41
5	The Shaping Machine	47
6	The Milling Machine	52
7	Chucks	57
8	Mandrels	66
9	Lathe Tools	70
10	Knurling	79
11	Lathe Operations (Miscellaneous)	84
12	Taper Turning	96
13	Lapping	102
14	Toolmakers Buttons	107
15	Milling in the Lathe	110
16	Dividing in the Lathe	121
17	Dividing	129
18	Drills and Drilling	140
19	Countersinking and Counterboring	155
20	Cutting Screw Threads	159
21	Cutting Screw Threads in the Lathe	170
22	Measuring Equipment	180
23	Marking Out	186
24	The Dial Indicator	202
25	Suds Equipment	209
26	Lathe Overhead Drives	212
27	Soldering, Brazing, & Case Hardening	217
28	Compressed Air in the Workshop	226
29	Some Additional Machine Tools	235
30	The Back Tool Post	241
31	Reamers	249

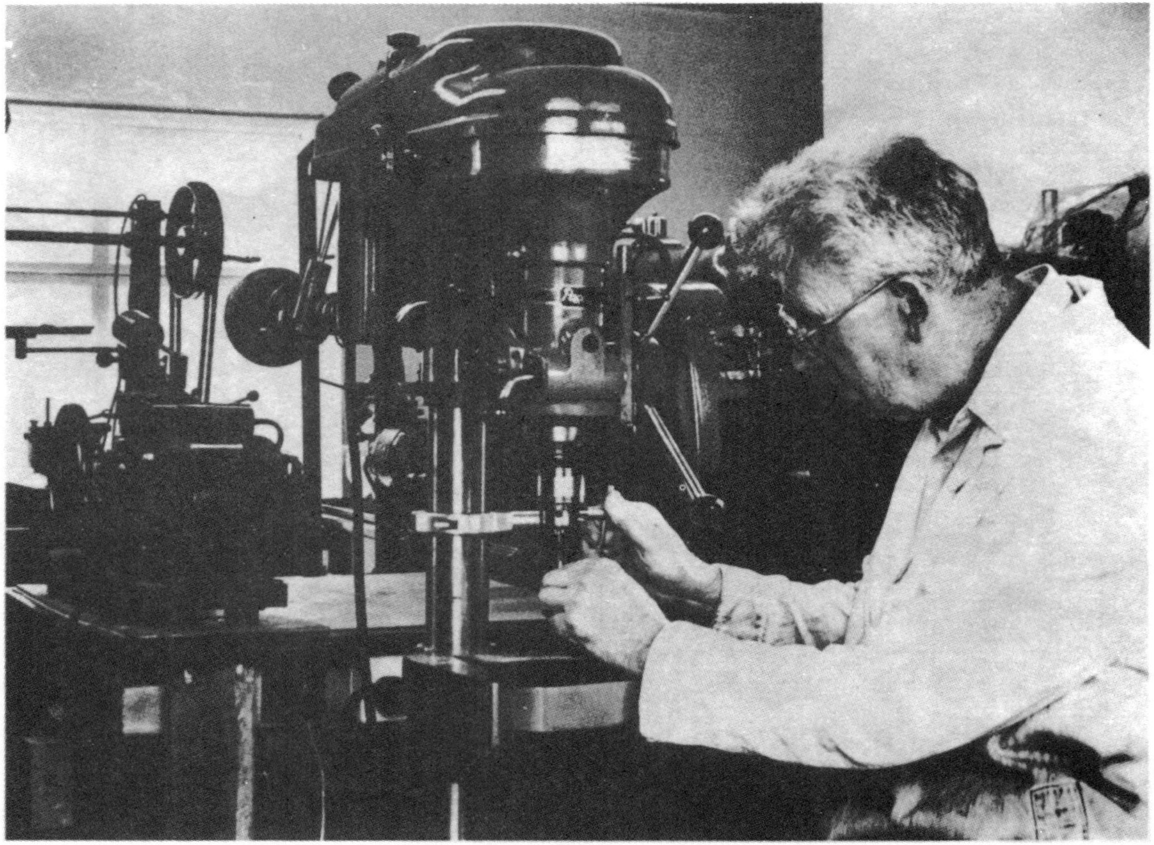

Fig. 1. The Author in his workshop

CHAPTER 1

The Workshop

THE Workshop itself may take many forms. It may be a room in the house, it may be an outbuilding attached to the house or it may be completely separate from it and constructed for the purpose.

Whatever form it takes, the workshop needs to be as well lighted as possible, both by natural and artificial light, and to have some means of keeping the shop dry and warm.

Where the workshop forms part of the house, the provision of artificial lighting presents little difficulty. The same may also be true of the heating system. But the natural lighting may well be inadequate.

For this reason all benching and machine tools should be located as close to the windows as practicable. In one workshop known to the author the workshop is an upstairs room having a large window embrasure with glass on three sides and the benching is located end-on to the main glass area. This enables the maximum of natural light to be concentrated upon the various machines grouped on the benches.

For those contemplating the planning of an amateur workshop on these lines, a warning must be given; make sure that the floor joists are strong enough to support the concentrated weight. Any reputable builder would advise on this point. Failure to make certain that the flooring is adequate could easily result in a disaster that would be most unlikely to win applause from the household authorities.

The Outdoor Workshop

Workshops that do not form part of the main house are either in outbuildings or in purpose-made wooden sheds of the garden type.

The outbuilding, where available, is probably the better of the two as the brick or stone construction makes the heating problem easier, moreover, if needed, the solid nature of the building forms an excellent anchorage for the benching.

The wooden workshop, however, presents more heating problems and, unless erected on a substantial concrete platform, is a difficult building to equip with benching and machine mounts.

The natural lighting, however, will present little difficulty since buildings of the garden shed type can be provided with plenty of window space. Moreover, the addition of roof lighting to a wooden building is quite an easy matter, though care must be taken to see that there is no water leakage from any skylight or transparent panel used for the purpose.

Artificial lighting for the separate building will, of course, be supplied from the electrical mains; and electricity may also, with advantage, be used to supply the heating.

The whole question of electricity in the workshop will be discussed later.

Workshop Furniture

Having decided on the most suitable type of workshop the amateur will have to consider the furniture needed for it.

Perhaps the most important item is the benching. It is not possible to buy really satisfactory benches so these are best made for oneself. For the most part the free-standing variety are the

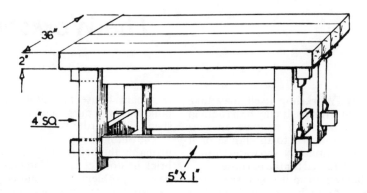

Fig. 2. A typical workshop bench

most practical as they need no support from the building itself.

They should be amply strong, which some commercial made benches are not, with legs made from at least 3 in. square material and having tops fashioned from 1½ in. deals, or better still 2 in. deals, seeing that they will very likely support a number of machine tools as well as forming the platform for mounting the vice used when sawing or filing.

A firm foundation for the vice is essential for it is near impossible to work either with accuracy or comfort unless the benching is rigid.

The illustration, *Fig. 2*, gives the essential measurements of a typical bench. (The height will, of course, depend on the stature of the person using it, and the size of the vice it is proposed to employ.)

A guide to vice mounting is so to arrange matters that the upper surface of the vice jaws is level with the point of the operators elbow when in the filing position. If the bench is also to be used for supporting machine tools, a comfortable working height for handling these will need to be considered and perhaps some compromise may have to be effected.

For the most part, however, the guide for setting the vice will serve to provide a comfortable working height for the machine tools themselves.

After deciding upon the size and location of the workbenches, thought must be given to the storage of tools and machine accessories as well as for the general engineering materials that will gradually accumulate.

Some may prefer to keep hand tools in racks, in which case these can easily be made for oneself. On the whole, however, tools are best kept in small chests of drawers while the machine accessories are housed in cupboards, preferably glass fronted so that they may be more easily identified.

Raw materials can be stored in wood or metal boxes of which at one time there was a plentiful supply in the form of surplus ammunition containers. These were of robust construction and one example at any rate might well form a model for making a number in wood. This container, shorn of some non-essential fittings, is illustrated in *Fig. 3*.

Nuts, bolts and screws are conveniently kept in old tobacco tins that have their ends painted with white emulsion paint so that the contents can be clearly indicated with one of the many laboratory marking pens now available. The tins themselves can then be placed conveniently in a shallow cupboard.

The number of chests of drawers and cupboards that are required will, of necessity, depend on the size and scope of the workshop. The author's experience is that, for the most part, it does not pay to make them oneself, for cheaper and perhaps better examples are to be had, often at a fraction of the cost in time and material, in one of the many auction sales that are now frequently held.

When choosing the furniture and deciding on other arrangements in the small outdoor shop, where space is naturally limited, remember to ensure that the entrance door opens **outward**. Failure to do so will lose valuable floor space that could otherwise be devoted to the siting of a cupboard or other storage medium.

Heating the Workshop

We now come to what is no doubt one of the real problems encountered when planning a small workshop, namely, how to heat it. The solution is of great importance, not only in relation to comfort when working there but also because of the valuable tools and machinery it will eventually contain.

Workshops forming part of the main house or attached to it will present little difficulty since the space devoted to them may well be already adequately heated, or at least can have the heating system extended to it. These remarks apply to household central heating systems as well as to electrical block heating with night storage, probably the easiest and most satisfactory method of them all.

On no account should the household heating furnace, if designed for solid fuel, be located in the shop. The attendant dust, inseparable from the stoking, is in every way such an unmitigated nuisance, and so damaging unless adequate precautions are taken, that it is scarcely possible to write about it in temperate terms. Happy the man, however, who has oil fired central heating; no problems of this sort arise for him.

Wherever the workshop may be located, on no account should any form of heating that discharges its products of combustion directly into the interior air be considered. This applies to gas or oil fired convector heaters and the like. These produce water vapour which, when the shop is cold, condenses on any unprotected

Fig. 3. A raw material storage container

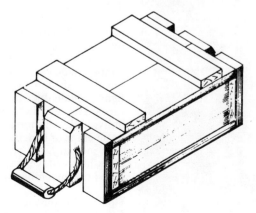

cold metal surfaces, and causes rust or corrosion.

The outside workshop, is of course, the most liable to the condensation nuisance. As a first step in getting rid of it the building needs to be lined, the space between the carcase and the lining being packed with an insulating medium. This will help to maintain an even temperature inside the shop and lessen the volume of the heating required to achieve it.

A workshop built in this way can be readily heated with a low-wattage electrical heater permanently left on or, if thought desirable, fitted with a room thermostat to control the system. The temperature range needed for both comfort and machine protection is from 60 to 65°F. so that the heat input needed is not large.

Electricity in the Workshop

It is now time to consider the whole question of the use of electricity in the workshop. For many years in the past gas was the medium employed for all heating, lighting and power requirements. In this connection many older readers will doubtless agree that a good gas engine in the shop possessed attributes of companionship never attainable by the electric motor. Be that as it may the versatility of electricity cannot be gainsaid, nor, for the most part, can its reliability be questioned.

Lighting

The artificial lighting of the workshop is a most important matter, the more so when it is the main source of illumination even in daylight.

The most satisfactory method is a mixture of general lighting supplemented by individual lamps where needed. For the former duty, both on the score of economy and efficiency, fluorescent strip lighting is to be preferred. Lamps and fittings for this purpose have nowadays been so simplified that apart from devising some simple way of hanging them up, and even this is not always necessary, they may be plugged into the mains much as an ordinary bulb would be.

Incandescent strip lamps are sometimes a convenience for lighting machine tools. An example is that shown in *Fig. 4* and *Fig. 5* where a lamp of this type with its reflector and support is seen attached to a Myford lathe. The maximum power of this class of lamp is 60 candles, generally considered sufficient for individual lighting.

High candle-power incandescent lamps can be used for general lighting but these do not provide the level of light obtainable from fluorescent tubes. They are therefore best kept for individually illuminating some specific machine or bench operation.

In industry, the practice of providing a low-voltage system for lighting individual machine tools is growing in

Fig. 4. Incandescent strip lighting fitted to a Myford lathe

Fig. 5. Incandescent strip lighting fitted to a Myford lathe

favour. The machines are safer to use when equipped with low-voltage lighting because the hazard of electric shock is completely eliminated, moreover, the smallness of the fittings needed allows the lights to be concentrated near the work without causing any obstruction to the operator.

The author has had a complete low-voltage lighting-and-power system in operation for many years and its convenience cannot be questioned. By means of this system it has been possible to make use of some of the many types of small electric motors produced for military purposes and to harness them to machine tool attachments or even to the machines themselves for duties such as the driving of lathe leadscrews when an independent feed was needed.

Power Supply

For many years now the lineshaft driven by a single prime mover has given place, in both the commercial as well as the private workshop, to individual electric motors fitted to the machine tools themselves. For the most part in the private shop these electric motors are of fractional horsepower and, as they are unlikely to be used all together, they can be connected directly to the single phase a.c. domestic mains.

In this connection it should be noted that the maximum individual loading permitted on the single-phase domestic supply is 1 h.p.

As an addition to the normal high voltage domestic power supply it is possible to use a low-voltage supply fed by a mains transformer or a transformer in combination with a rectifier when a d.c. supply is needed. These matters will be dealt with in detail later.

Plug Points

There should be plenty of electrical plug points in the workshop. This will simplify connecting up the various machines and any special lighting required. In addition if one or two plug points are placed at convenient locations below the benches such equipment as electric hand drills and soldering irons can be brought into use. The sockets used can either be plastic or iron-clad; but the plugs should be of the rubber kind. These will withstand the hard knocks they are likely to receive in service.

The sockets must be correctly connected into the domestic system so that they are properly earthed. With advantage this may be work for a qualified electrical contractor, especially if there is a number of points to be arranged. He will have all the necessary equipment to check his work and ensure that it is satisfactory.

Cables

The electrical cables used in the shop

Fig. 6. Cable support

should be of a tough variety able to withstand abrasion. This requirement applies to both lighting and power cables. These should not be allowed to trail about the floor but be supported above head height so as to cause no obstruction.

Cables about the workshop are inevitable. For ourselves we have devised a simple method of suspending these by means of an adjustable support attached to the roof by a simple hook. This fitment is illustrated in Fig. 6.

It consists of a hanger that can be suspended by a hook from any convenient point in the roof. On the hanger is placed a crutch that can be adjusted for height and then locked. The cable or cables are then dropped into the crook of the crutch and held well out of reach.

The size and type of cables required will, of course, depend upon the electrical loading they have to carry. Those for lighting will have a maximum carrying capacity of 5 amperes and cables for power 15 amperes with a third core enabling any apparatus to be properly earthed.

In the matter of low-voltage lighting the cables for this purpose should have a maximum capacity of 5 amps., this will be sufficient to light the single lamps normally available. Therefore, the cables suitable for mains lighting will serve.

Low voltage power cables, however, will need to have considerable current carrying power. It is absolutely necessary that the current should be fed to motors without any drop in voltage or their power output will be much diminished. As an example of this requirement, motors used in the author's workshop can be cited. Some of these normally operate at 20 volts a.c. and are rated at $\frac{1}{3}$ h.p. Therefore, for the moment disregarding any figure for the efficiency of the motors themselves, in the first instance the current needed will be of the order

$$\frac{746 \text{ watts}}{3 \times 20 \text{ volts}} = \frac{248 \text{ watts}}{20 \text{ volts}} = 12 \text{ amps.}$$

approx.

The efficiency of these low-voltage motors, for practical purposes can be assumed at 50 per cent. Therefore the current needed will be twice the theoretical figure, say 25 amps. To carry such a loading 30 amp. cables could suffice, these are often obtainable from suppliers specialising in the sale of surplus equipment.

When a correct-sized cable is used there should be no sign of it warming up. If a cable does get warm then there will be a drop in the supply voltage. If the cable gets **hot** then the volts drop is considerable and the cable is unsuitable.

In order to accommodate cables of the substantial cross-section required the plugs and sockets will need to match. In days gone by the two-pin power plugs, and their attendant sockets, were of quite heavy proportions.

Our experience is that fittings of this type sometimes appear amongst the early lots in auction sales where, because there is usually no competition to secure them, they may be bought for a fraction of their original cost.

The use of modern three-pin plugs for duties with a low-voltage system is not advised. It could be fatally easy to plug one of these, attached to some piece of low-voltage equipment, into the domestic supply, and the results might be disastrous.

Low-voltage Power Supply

The low-voltage supply may be obtained from the domestic electricity mains in a variety of ways, in accordance with whether it is an alternating current, a direct current supply or combination of both these is needed.

The alternating current supply will be obtained from a transformer and this may be one of two types. The first is the common transformer illustrated diagrammatically in *Fig. 7*.

This piece of equipment consists essentially of a laminated soft iron core upon which there are two distinct and independent windings. The first of these, the Primary Winding, marked 'P' is connected to the domestic supply whilst the Secondary winding 'S' sometimes 'tapped', as it is termed, to provide an alternative voltage, is connected to any low-voltage apparatus it is needed to supply.

This type of transformer has the great asset that it is most unlikely to provide the user with an electric shock. There is no electrical connection between the Primary and Secondary Winding so the chance of a high-voltage component being fed to the low-voltage side of the system is most unlikely.

The other type of transformer we have to consider is that known as the auto transformer depicted diagrammatically in *Fig. 8*.

This illustration represents a typical example as used by the authors. As will be seen there is but a single winding with tappings supplying voltages of the values indicated. It will be obvious that transformers of this type will need to be properly earthed, for it must be remembered that, here, there is no isolated primary winding so the risk of electric shock is very real. The output from the transformer is taken connecting from the zero tapping to which the neutral (black) wire from the mains must be connected.

Rectifiers

If a low-voltage direct-current supply is needed this is best met by the

Fig. 7. Common transformer

Fig. 8. Auto transformer

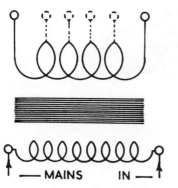
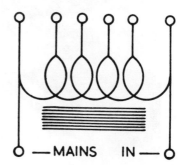

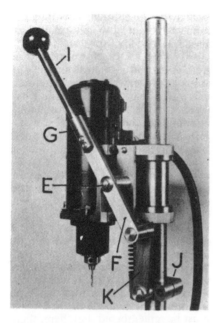

Fig. 9. Electric drill

addition of a rectifier connected to the output of the transformer. Modern techniques have produced apparatus capable of rectifying quite large currents. They are very reliable and their manufacturers are always pleased to supply details and suggestions as to the most practical ways of using them.

Another way to procure a low-voltage electricity supply is to make use of a motor-generator set run from the mains and provided with a regulator in the field circuit of the dynamo to control its output. In this way the generator, which of course runs at a constant speed, can be set to deliver an output suited to the work in hand.

Our own version of the motor-generator set consists of a ⅓ h.p. a.c. motor direct coupled to a d.c. generator once forming part of the lighting system of a somewhat historic motor car, the 30-98 Vauxhall circa.1922.

The equipment is mounted on a light four-wheel trolley enabling it to be taken about the workshop or indeed anywhere else its presence might be needed.

It is no bad thing to have the means of low-voltage supply in a portable condition as it greatly increases its usefulness. This also applies to the d.c. supply obtained from a metal rectifier. In our case the transformer that feeds it is in a wooden box provided with a substantial carrying handle, whilst the rectifier itself is fastened to the outside under a ven-

Fig. 10. Spot drilling attachment

tilated protective cover. The transformer will give an output of some 700 watts whilst the rectifier can handle about 200 watts at 20 V d.c. comfortably.

Readers may wonder what are the applications of a low-voltage supply in the workshop, apart from the obvious one of local lighting. They may rest assured that the applications are numerous.

In the first place for example, motors fed from this supply can be used to drive independently the lathe leadscrew or the feed screw of the cross-slide. This facility is essential to some of the operations when milling in the lathe in particular on the occasions when the headstock itself is holding the work and therefore stationary.

Low-voltage motors can be controlled simply by connecting a variable resistance in series with their supply, therefore they may conveniently be used to drive a milling spindle mounted on the lathe saddle or a spot-drilling attachment similarly located.

In addition the motors may be used to power toolpost grinding spindles, cutter grinders, drills both hand and portable, and rotary filing equipment to name but a few applications. Some of these are illustrated in *Figs. 9, 10 and 11.*

Suitable Motors

The motors that are suited to these various duties can be divided into two categories, the high and the low-power. The high-power motors, giving about $\frac{1}{3}$ h.p. are capable of driving milling spindles, toolpost grinders, cutter grinders and hand drills; whilst the low-voltage class, consuming about 40 to 50 watts are very suitable for use with independent feed screw drives and spot drilling attachments for the lathe where a small motor is essential on account of space and light weight.

Motors of both types are often found on the surplus market where they have gravitated having been removed from various pieces of redundant military equipment; as might be expected they are of an exceptionally high standard.

There is of course little demand for them so they may often be bought for a small fraction of their original cost.

Provided that the input current to these motors is low they will cool themselves automatically. Motors requiring a high input current, however, will need to be artificially cooled par-

Fig. 11. Cutter grinder

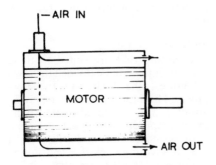

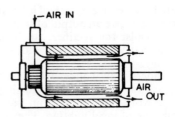

Fig. 12. Air OVER frame cooling

Fig. 13. Air THROUGH frame cooling

ticularly if they are to be used for extended duty; a purpose for which they were not originally intended.

Cooling of these motors can be achieved in two ways, both needing a separate small motor driven blower of a type sometimes used for cooling radio transmitter valves and also to be obtained on the surplus market.

The simplest method of the two is to blow air **over** the outside of the motor. A typical set-up for carrying this out is seen in the illustration *Fig. 12*. Here the motor is surrounded by a metal case, made from a pair of coffee tins and connected to the blower output. The case or sleeve is open at one end whilst the air input is at the opposite end. In this way the coolant is directed along the length of the motor.

The alternative method is more difficult as it involves stripping down the motor and drilling a ring of holes in the drive-end bearing plate. The air from the blower is then taken to a connection on the detachable cover over the brushes and air is forced **through** the inside of the motor to cool it. Both methods are illustrated diagrammatically in *Fig. 13*.

Motor Internal Connections

For the most part an electric motor consists of two main elements. (1) the Armature (or Rotor) and (2) the Field Magnet. When the motor is intended for use with the d.c. supply these elements are connected together in one of two ways.

In the first of these the Armature and field are connected in the manner seen in the illustration *Fig. 14* at A. This is known as a **shunt** field. When the connections are made in the way illustrated at B it is called a **series** field.

Of the motors we have been considering those of low power are connected internally as shown at (A) whilst those capable of considerable

Fig. 14A. Shunt wound motor

Fig. 14B. Series wound motor

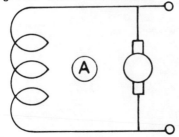

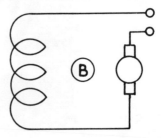

THE WORKSHOP

power output for the most part have their internal connections made in the manner shown at (B). The latter arrangement will also enable the motors to be used with an alternating current supply.

Small alternating current machines, even if available, are really not a practical proposition. They possess little power and lack flexibility.

Switching for Small Power Motors

Before we leave the subject of low-voltage electricity a word must be said about the switches that will be needed to control the direct current motors employed.

For the small power machines any well made switch with a 5 amp. rating will serve. In addition some form of switch, enabling the motors to be reversed, will be needed. Here, again, the surplus market can often provide a solution to the problem. A wiring diagram showing a typical arrangement is given in *Fig. 15*.

This illustration depicts a change-over switch having its centre poles connected to the armature of the motor to be reversed, whilst its field winding is cross-connected to the outer poles of the switch. In this way, simply by moving the contact blades to one side or other of the outer poles, the field winding connections are reversed in relation to the armature, thus altering the direction of the motor's rotation.

Heavy Current Switches

Switches for use with low-voltage motors having considerable power output need to be capable of handling heavy currents. There appears to be nothing on the surplus market that will satisfy the requirement in an uncomplicated way so, if a neat and compact switch is wanted, we must

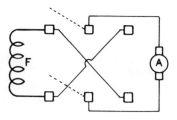

Fig. 15. Reversing switch

make it for ourselves. The basis of a suitable switch is the design used in many of the starter switches fitted to motor cars. A diagrammatic representation of the arrangement is given in the illustration *Fig. 16*.

A switch of this nature consists firstly of an insulated base carrying a pair of brass quadrants B_1 and B_2. It is further provided with two concentric serrated brass cups C_1 and C_2, mounted on a plunger D. When this is depressed the two cups make contact with the brass quadrants thus connecting them together electrically. The circuit is broken when the cups are separated from the quadrants by the return spring seen surrounding the plunger.

As a result of the large area of contact provided by the concentric brass cups, a switch of this type will handle heavy currents without any detrimental arcing. Moreover, the design pro-

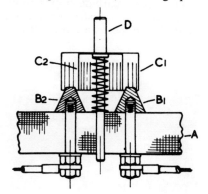

Fig. 16. Heavy current switch

Fig. 17. Foot switch for transformer

vides a complete self-cleaning system and is particularly useful in a highly-inductive circuit when, for example, it is used in the primary circuit of a transformer. The foot switch illustrated in *Fig. 17* was designed and made for just such an application. The same type of switch is fitted in the handle of the electric drill previously illustrated. The details of this device are illustrated in *Fig. 18*.

The Bench Grinder

As has so often been pointed out, some means of grinding tools is essential in the metal-working shop. Without it no lathe tools or drills can be sharpened nor is it possible to make a creditable job of re-shaping screwdrivers of the instrument type, as for the best results these tools need to be hollow ground.

The process of sharpening tools on the wheel is known as 'off-hand grinding', and it is a curious fact that, except for treating specialist equipment such as 'carbide tipped tools', the practice in industry really is off-hand, for no means are provided, on the grinding machine itself, of ensuring that the correct angles are imparted to the tools.

Fortunately, the small workshop can remedy this deficiency for itself by fitting to the bench grinder a tool rest capable of being tilted at an angle in relation to the side face of the grinding wheel.

The illustration *Fig. 21* depicts such an angular rest fitted to a small grinder intended for the treatment of small turning tools. The make-up of this device will be almost self-evident consisting as it does of but three simple parts, the rest table, the support, and a bracket for attachment to the base of the machine.

The use of the device is of course restricted to grinding on the side of the wheel, for only in this way can the correct angles be imparted to the tool face.

The illustration *Fig. 19* shows a similar angular rest fitted to a commercial electric grinder.

The parts of the device are similar to those shown in the previous illustration, and the method of adjusting the tool angle to be ground is the same.

It is of course, perfectly possible to make an adjustable rest that will allow the periphery of the wheel to be used for grinding. But the small wheels normally used by amateurs tend to produce a hollow-ground finish and weaken the cutting edge of the tool itself. For this reason grinding on the side of the wheel is to be preferred.

Fig. 18. Parts of heavy current switch fitted to low-volt electric drill

THE WORKSHOP

Fig. 19. An angular rest fitted to a commercial grinder

If it is possible, two grinding wheels should be available in the shop. One, a wheel of 60 grit size for rough shaping the tool, the other an 80 grit wheel to be used for finish grinding the tool point.

When the grinder is of a reputable make it is safe to assume that the grinding wheels themselves will be properly mounted. It is worthwhile, nevertheless, to assure oneself on this point. A correctly mounted wheel is depicted diagrammatically in *Fig. 20*.

The wheel, having a lead bushing at its centre, is gripped between a pair of dished flanges. One of these is a firm fit, and is sometimes keyed, to the spindle. The flange diameters are normally one-third of that of the grinding wheels themselves and paper washers are interposed between the flanges and the face of the wheel to ensure that it beds properly. The flanges must never contact the wheel directly and to ensure they do not, grinding wheel manufacturers supply the washers already affixed to the wheel.

The wheel itself must run without any wobble. If there is any, the fixed flange should be examined and its contact face checked for true running. Wobble at this point, provided the fixed flange is secure, can sometimes be corrected by turning the flange in place. But this, admittedly, is not an easy job so it is better to withdraw the flange, mount it on a true running mandrel on the lathe and machine the part in this way.

The grinding wheel must be run at the correct speed of approximately 5,000 surface feet per minute in accordance with the following table:

Wheel diameter	Revs. per minute
3	6,400
4	4,800
5	3,800
6	3,200

In use, the individual grits of which the wheel is composed become dulled and do not cut freely. It is necessary, therefore, to dress the wheel from time-to-time in order to remove the worn grit and present a new cutting surface. This is carried out with a wheel dresser, a tool available in most tool shops. The operation is accompanied by a considerable amount of dust, naturally of an abrasive nature, so the siting of the bench grinder in the workshop is of some importance.

Whenever possible, grinding machines should be placed well away from other machines. If this cannot be done the tools must be covered to protect them during the wheel turning process.

Grinding Screwdrivers

Screwdrivers are not the easiest tools to re-shape free hand. Instrument screwdrivers, in particular, need careful treatment and should, for preference, be hollow-ground. Provided the screwdriver to be ground is of a medium size the set-up illustrated in *Fig. 21* can be used. A small diameter

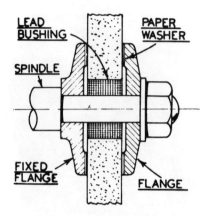

Fig. 20. A correctly mounted grinding wheel

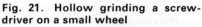

Fig. 21. Hollow grinding a screwdriver on a small wheel

wheel is needed mounted directly on the shaft of the driving motor or, if need be, on a separate spindle. In order to make sure that the grinding is symmetrical, the shank of the tool is painted with blue marking and a ring scribed round it. The ring is brought into contact with the edge of the grinding rest each time the tool is turned over; in this way the ground faces will be symmetrical.

A more sophisticated device is depicted in *Fig. 22*. Here the marked screwdriver, resting against the edge of the grinding rest, is replaced by an adjustable angular fence consisting of a table upon which the fence is mounted and a rail to accept V-grooves machined in a block to hold the blade of the screwdriver. This enables grinding to take place either upon the periphery of the wheel where instrument screwdrivers need to be serviced or on the side of the wheel in the case of standard drivers.

The use of the angular grinding rest is again dealt with in some detail later in the book where, in Chapter 9, the servicing of the lathe tools is being considered.

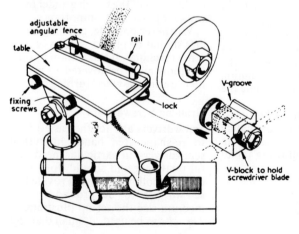

Fig. 22. Fixture for grinding screwdrivers

CHAPTER 2

The Lathe

THE lathe is the most important single item in the equipment of the amateurs workshop. With it, when fitted with the necessary attachments, we can solve almost any problem in machining that may be encountered. Many of these additional fixtures may be purchased as commercial products whilst those of a specialised nature are not difficult to make using the lathe itself to help in machining the necessary parts; specialist fixtures will be dealt with in a later chapter.

The type of lathe most suitable to the requirements of the amateur is one of simple but robust design having facilities for the adaption of a range of commercial attachments produced especially for it as well as for the specialist fittings already referred to. The production of such a tool at a price acceptable to a wide range of buyers eventually entails manufacture on the large scale, and nowhere is this better exemplified than by the products of the Myford Engineering Company whose lathes have, for many years, achieved a world-wide reputation.

Whilst there are some amateurs whose financial resources enable them to buy lathes having the maximum of refinement, for the most part purchasers in the amateur field do not need such equipment, requiring only the basic machine tool to which simple attachments may be added from time-to-time. For this reason it is the simple lathe that will be the subject of the present treatment.

The Drummond Lathe

However, before dealing with lathes specifically designed and manufactured today for amateur use, mention must be made of their most important predecessor. This was the Drummond Lathe of 3½ in. centre height. Made specifically for light machining such as would be encountered in the small

Fig. 1. The Drummond 3½″ lathe

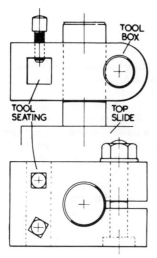

Fig. 2. The Drummond toolpost

workshop, the lathe had, or has for there are many still in existence, a number of outstanding features and some of these may be seen in the illustration *Fig. 1*.

The first, and perhaps the most important, of these features is the lathe bed itself. As will be seen the casting is of cantilever or anvil form having a single mounting foot of large area set nearly at this casting mid point. This ensures that, when bolted down to a bench, there is no possibility of disturbing the lathe bed and destroying its accuracy.

The second feature is the somewhat original toolpost fitted to the top slide of the Drummond lathe. This enables any tool in the toolbox to be quickly set at centre height without resorting to the collection of packing often associated with this operation.

The Drummond toolpost, illustrated in *Fig. 2* consists of a toolbox clamped to a pillar integral with the top slide. After clamping in the tool seating the tool point is readily set at both correct height and position by slackening the clamp bolt and raising or lowering the toolbox as required.

There are many more features that made the Drummond lathe, when in production, one of the most successful ever produced for amateur use; not only the robustness of the saddle or slides as well as the bed itself but space does not allow of their description here.

In England the immediate successors to the Drummond lathe in the light engineering field were the range of machines produced by the Myford Engineering Company. These have now developed into two main products; the ML10, a back geared screw cutting lathe of $3\frac{1}{4}$ in. centre height, admitting 13 in. between centres and the ML7 and its derivatives lathes of $3\frac{1}{2}$ in. centre height admitting 19, 20 and 32 in. between centres. There is also a wide range of accessories and equipment available from the manufacturers.

Standard Equipment

All screw-cutting lathes are supplied with a set of standard equipment. This usually consists of a set of changewheels for screw cutting, a face plate, a catch plate for driving work mounted between centres, a pair of centres and the necessary driving belts and their guards.

In this form the lathe as supplied is only suitable for mounting on the bench and the work it will handle restricted to operations between centres or on the faceplate.

It follows, therefore, that some additional fitments are needed to enable a normal range of machining operations to be undertaken. Apart from some means of driving the lathe, and today this infers an electric motor for which a mounting, together with the necessary built-in countershaft, is normally provided, chucks will be

THE LATHE

Fig. 3. The 4-tool turret

needed so that both bar material and castings can be machined. In addition a special chuck to be mounted in the tailstock mandrel will be required in order to grip drills and centre drills, taps and other tools.

All these matters have been dealt with in many lathe books, so it is only necessary to mention them in outline here.

Toolposts

The standard toolpost supplied with the lathe is one admitting a single tool only. This is a somewhat restrictive arrangement that can be largely overcome by substituting the 4-tool turret seen in the illustration *Fig. 3*. This accessory is normally a manufacturers supply and is provided with a simple means of indexing to ensure that tools will take up their correct station after the turret has been rotated to bring them successively to bear on the work.

The turret is designed to admit ground steel toolbits, in the case of the Myford lathe, up to ⅜ in. section.

In conjunction with the 4-tool turret, or indeed in its absence, the Back Toolpost illustrated in *Fig. 4* is a useful adjunct since it enables a parting tool to be permanently mounted at the rear of the cross-slide and rapidly brought into use, a facility of great value when machining small numbers of similar components.

Back Toolposts made for light lathes usually have a single bolt fixing and have means of tool height adjustment either by packing or by a radially seated 'boat' that supports the tool and allows it to be adjusted with greater ease.

The Back Toolpost has further advantages if it can be provided with its own turret to hold an additional tool, such as one for chamfering; the design for a Back Toolpost to hold two tools will be discussed in a later chapter.

Steadies

Steadies to support long and slender work mounted between centres, or projecting some distance from the chuck, can be supplied by the lathe makers but only as additional attachments; the serious user cannot afford to be without them. There are two forms of steady, the first illustrated in *Fig. 5* is known as a fixed steady, and is used to support work projecting from the lathe headstock. It is fastened to the lathe bed along which it can be

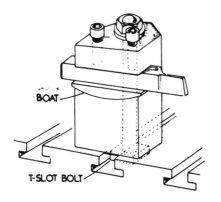

Fig. 4. The back toolpost

Fig. 5. The fixed steady

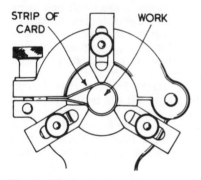

Fig. 7. Method of avoiding damage from the steady jaws

Fig. 6. The travelling steady

turned and be secured in the position desired.

The second or Travelling Steady, seen in *Fig. 6*, is attached to the saddle of the lathe and is brought to bear on work mounted between centres. It enables the turning tool to be set close to its jaws and prevents the work itself from backing away from the toolpoint The steady has a pair of jaws, restraining the work in both the horizontal and vertical direction, and these, in common with the jaws of the fixed steady, are usually made of bronze. The jaws are, of course, adjustable and can be clamped where required in relation to the work.

Metal jaws bearing directly on relatively soft material such as aluminium alloy are not to be recommended since they tend to score the work. But this difficulty can be overcome if a strip of well-oiled card is interposed between the jaws and the work and secured against rotation by being clamped, in the case of the fixed steady, between the upper and lower halves of the attachment. This is a tip that seems unknown to a great number of turners so that no apologies are offered for bringing it to readers notice. The method of holding the card is illustrated in the diagram *Fig. 7*.

This method of avoiding damage to the work has been used by us many times in the past and has enabled work to pass the somewhat exacting requirements as to finish imposed by such bodies as the atomic energy authorities.

Centres

Part of the standard equipment provided with a lathe is a pair of centres. These fittings are a survival from the earliest days when practically all turning was carried out between centres. Of those supplied, one is hardened, the other left soft. The former fits the tailstock whilst the latter is set in the headstock mandrel and can be turned

THE LATHE

Fig. 8. Half-centre and female-centre

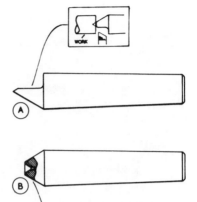

in place to make certain that any work mounted between the centres, will run true.

Additional hardened centres available are half-centres, enabling slender spindles to be turned without the lathe tool fouling the toolstock centre itself, and female centres to accommodate pointed spindles.

These centres are illustrated in *Fig. 8* at (A) and (B) respectively. Female centres are principally of interest to clockmakers and instrument workers whose activities sometimes include work on pointed shafts.

One other form of centre deserves notice if only for its antiquarian association. This is the pump centre illustrated in *Fig. 9*. It may be used to centre work that has to be mounted on the lathe faceplate when it is essential that the part to be machined is held symmetrically.

The pump centre consists of a tapered body (A) designed to fit into the lathe headstock mandrel, a centre (C) that is an accurate sliding fit in the body, and a compression spring (E) bearing on the centre (D) and held in place by the cap (B).

An example of the use of this form of centre is in the setting of a trunk piston casting in the lathe so that the seatings for the gudgeon pin may be bored.

Here, after the casting has been machined all over to the correct dimensions, the work is marked off for the location of the gudgeon pins holes. These centres are then lightly spotted

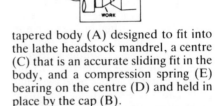

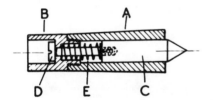

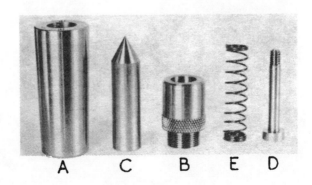

Fig. 9 and Fig. 9A. The pump centre

THE AMATEUR'S WORKSHOP

Fig. 10. The tumbler gear and Norton gearbox

with a centre drill. The faceplate is now mounted on the lathe mandrel nose and the pump centre inserted in the mandrel bore. The Tailstock with centre in place is then brought up and the work gripped between the two centres. It is then a simple matter to offer up an angle plate fixed to the faceplate and to secure both the angle plate and the work in perfect alignment. When this has been carried out the faceplate is temporarily removed

Fig. 11. The ML10 lathe

to allow the pump centre to be withdrawn.

In addition to the centres described some manufacturers can supply, as additional fitments, prong centres enabling the turning of large wooden objects to be undertaken, and adaptors carrying drill pads for use in the lathe tailstock. These drill pads will be described in a later chapter.

The Leadscrew

As might be expected, in lathes made at a competitive price, the Leadscrew serves to drive the saddle along the lathe bed both for normal turning purposes as well as for screw-cutting operations. The Leadscrew itself is driven from the headstock mandrel through a train of pick-off gears or change wheels that may be varied to accommodate the rate of feed required or the pitch of the screw it is desired to cut. In addition a device known as the tumbler gear allows the drive to be reversed when needed.

THE LATHE

Fig. 12. The ML10 lathe on stand

An alternative to the change wheels is a gearbox allowing a wide range of feeds and thread pitches to be selected simply, for the most part, by the movement of two levers. This arrangement is seen in the illustration *Fig. 10* where the tumbler gear and its operating lever is also depicted.

The Myford Range of Lathes

The range of Myford lathes has already been mentioned, but the time is now opportune to describe these in greater detail. The smallest lathe the ML10 is the latest production of the Myford Engineering Company. In order to keep the price to one attractive for the amateur and at the same time provide a thoroughly practical machine tool, the specification of the lathe has been pruned of any item that would otherwise raise the price of the basic machine to an unacceptable level.

The lathe is supplied in the first instance as a bench machine with the usual standard equipment already referred to. But there is an extremely wide range of accessories available, for the most part interchangeable with those applicable to the ML7 lathe.

No tumbler gear is fitted and the back gear is engaged by the simple expedient of providing a fork, cast on behind the front bearing, to house the back gear cluster, allowing the gears to be slid into engagement and then locked to position. The same arrangement is used to provide a means of reversing the Leadscrew drive. These are but two examples of the simplification already referred to.

A further example is the omission of the rack for the quick-traverse arrangements for the saddle. In this instance the pinion of the traverse handle engages the leadscrew directly.

The headstock spindle is hardened and ground and runs directly in the iron casting of the headstock, an arrangement that should provide bearings having long working life if

Fig. 13. The ML7 lathe with gearbox

experience with cast iron bearings, extending over many years, is any guide. The spindle nose in accord with other machines in the Myford range is bored No. 2 Morse Taper.

The tailstock, having a rigid and robust casting, is adjustable so that it may be set over for taper turning and has a spindle also bored No. 2 Morse Taper.

The lathe is provided with an electric motor platform at the rear and the drive to both the countershaft and the headstock is by V-belt.

Supplied in the first instance as a bench machine and illustrated in *Fig. 11*, the lathe can also be provided with the stand seen in the illustration *Fig. 12*.

The Myford ML7 Lathe

The most important of the Myford Engineering Company's productions is the range of machines and accessories based on their ML7 lathe. One example of this tool is illustrated in *Fig. 13* though not in its basic form.

The simplest machine is one having a specification not unlike that of the ML10 already described, but with the addition of lever operated back-gear and tumbler-gear assemblies. The Headstock and Tailstock spindles are both bored No. 2 Morse Taper and the former is carried on white metal bearings provided with pick-off shims for adjustment purposes. This adjustment is carried out in the first instance, at the works. Provided the lathe has proper treatment our experience is that despite many hours work no further adjustment is required even after 30 years use.

In the case of the ML7 the pinion of the quick traverse gear engages a rack set under the ways on the front face of the lathe bed casting.

The top slide is completely detachable from the cross slide leaving the latter quite free for the mounting of the many accessories that are available, or for use as a boring table.

The Leadscrew is normally driven through pick-off gears and is reversible, as has already been mentioned, by a tumbler gear system placed at the rear end of the headstock casting. For those who need the facility, a gearbox

THE LATHE

is available for driving the leadscrew, thus enabling rapid selection of thread pitches and feeds to be made. The lathe so fitted is illustrated in *Fig. 13*.

The Myford Super 7 Lathe

The Super 7 lathe is based on the design of the ML7. Its main difference from the latter is in the headstock mandrel bearing arrangements, and the configuration of the mandrel itself.

A section of the Super 7 headstock is shown in the illustration *Fig. 14*. The front end of the spindle runs in a plain cone bearing whilst the rear end is supported by a pair of angular contact races adjustable to take up shake in both journals as well as any end float in the spindle itself.

Levelling the Lathe

No machine tool can function properly unless it is set level before being put to work. This is particularly true of the lathe and never more so than when the tool is to be mounted on a wooden bench.

The cantilever form of bed casting, of the type used by the Drummond lathe, is the most suitable for mounting on a wooden bench in fact it is the only practical trouble-free shape.

Lathe beds with separated mounting feet are really not suitable for setting on wooden benches because of their liability to cause distortion of the bed. In fact, most manufacturers ask purchasers to avoid placing their lathes on a wooden foundation.

For the most part lathes of the size most suitable to the amateur are best placed on a cabinet stand. These are of all-metal construction so, provided that the stand itself is on a solid foundation, once the lathe has been correctly levelled on the stand, accuracy will be maintained.

In large organisations employing many machine tools the millwrights who are responsible for setting new tools in place, have sensitive spirit levels and other instruments to ensure that machines are correctly levelled.

The amateur usually is not so finely equipped but he can still achieve success with the facilities he possesses.

A dial test indicator is normally available in the small workshop and it is this instrument that can be used to assist in the levelling operations. The lathe is first mounted on the bench or

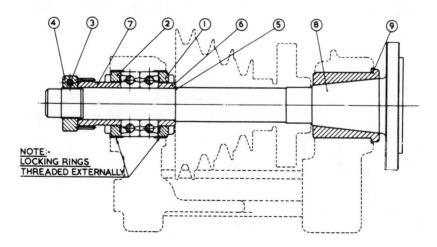

Fig. 14. ML7 Super 7 headstock in section

cabinet stand with its holding-down bolts in place but not secured. A piece of round bar material, about 1 in. diameter and some 10 in. long, is then gripped in the chuck and set to stand out about 8 in. from the jaws.

The dial test indicator is then brought into contact with the bar and maximum and minimum readings are taken by rotating the chuck by hand. The mean of the two readings is next established again by turning the chuck, and the indicator set to zero at the mean position.

If the holding-down bolts are now tightened any distortion of the lathe bed will be indicated by movement of the indicator needle. Shim packing will then need to be placed under the feet of the lathe bed until the indicator reads zero when the bolts are fully secured.

Having levelled the bed in the way described which accords with the makers instructions, a check must be made on the accuracy of the setting. The procedure is illustrated in the diagram *Fig. 15*. A test piece, about 1 in. diameter and some 6 in. long is turned and relieved in the manner shown in order to provide two collars about $\frac{1}{2}$ in. wide. These are each machined with the same tool setting taking a cut of about 0·002 in., the test piece being unsupported by the tailstock. A micrometer is then applied to each collar in turn, and its reading noted. If both readings are identical no further action is needed. If on the other hand there is a discrepancy this indicates that further adjustment to the packing of the foot at the tailstock end is required. Should this be so, and the collar on the outer end of the test be the larger, the packing must be placed under the **front** of the foot. If the collar is smaller insert it under the **rear** of the foot. Readers should note that two or three checks may be needed to secure final accuracy.

Caring for the Lathe

Earlier in the chapter reference has been made to the proper use of the lathe. Readers may well think this is a matter that is not only self-evident but scarcely worthy of comment. Experience has shown however, that, in the main, the treatment of machine tools leaves much to be desired so no apology is offered either for referring to the matter or for suggesting some practical ways of caring for a lathe.

In the first place all standing machinery should be covered when not in use; this too has already been stated. The atmosphere generally is

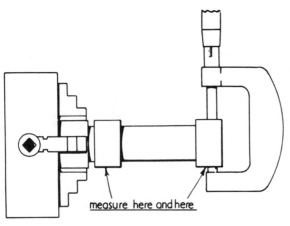

Fig. 15. Checking the accuracy of levelling the lathe bed

THE LATHE

full of dust and the air in the workshop particularly so; therefore, it is important to take any reasonable steps to reduce the deposit of this possible abrasive substance on the tools themselves.

It goes without saying that a lathe needs to be regularly lubricated. The makers are very definite about this and for the most part issue precise instructions on the matter. It is most important that the mandrel bearings are replenished with oil each time the lathe is used for an extensive run, and that the sleeve bearing of the driving pulley is lubricated as laid down by the manufacturers. This also applies to the back gear assembly.

Many of the other bearings in the lathe are provided with oil retaining bronze bushes, so it is only necessary to lubricate these at infrequent intervals as laid down by the manufacturers.

It is also important to make sure that swarf, that is the collection of metal particles which are the products of turning operations, is kept away, so far as is possible, from the surface of the lathe bed and slides as well as from the Leadscrews and the feed screws.

The feed screws are usually protected by their own slides, whilst the manufacturers provide some form of clip guard for the leadscrew.

But the safeguarding of the lathe bed is a matter for the user himself. A chip tray attached to the cross slide is the best solution to the problem, the device being on the lines of that shown in *Fig. 16* and *Fig. 17*.

In addition to the chip trays for the protection of the saddle the lathe should have some means of preventing swarf from dropping on to and lodging in the leadscrew. Many lathes are so provided, but for those who do not already have them the fitment illustrated in *Fig. 18* is quite easily made from a short length of cycle tubing affixed to a lug enabling it to be attached to the lathe saddle. As illustrated the Leadscrew guard is suitable for attachment to the Drummond lathe and this applies also to the chip trays seen in previous illustrations.

The trays may be affixed to the saddle in many different ways, but, because quick detachability for clean-

Fig. 16. The chip tray—ML7

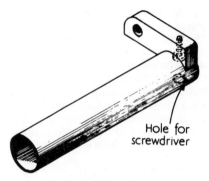

Fig. 18. Leadscrew protector

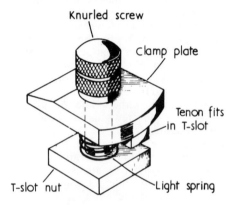

Fig. 19. Clamp for chip tray

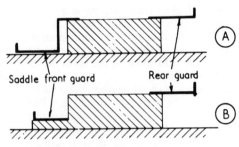

Fig. 20. Arrangement of chip trays for the Drummond

ing or other reasons is important, the clamp seen in the illustration *Fig. 19* is the best. The use of these devices, fitting the T-slots as they do, enable chip trays to be put on or taken off in a minimum of time. They were designed for one-handed operation and are self-releasing by virtue of the light spring that pushes up the clamp plate as soon as the knurled screw is turned.

Possibly alternative placing of the chip trays is depicted diagrammatically in the illustration *Fig. 20*. The arrangement for the Drummond lathe is outlined at A whilst the lay-out for the Myford lathe is seen at B. A

Fig. 17. The chip tray— Drummond

similar layout is also suitable for any other lathe having wings to the saddle. Chip trays may be extended along the axis of the lathe but care must be taken to avoid contact with the chuck.

Finally a word of warning must be uttered on the subject of overloading the machine. Probably the greatest damage that can be done to the mandrel of a light lathe is brought about by knurling. This is an operation involving the use of a single-wheeled tool, for the most part fed radially into the work. As may be imagined this imposes a considerable strain on the mandrel and its bearings. The loading can be somewhat reduced, however, if the tailstock is brought up to support the work. A better form of knurl wheel holder is one having two wheels diametrically opposite that embrace the work to it. This tool relieves the mandrel of all bearing loading; it will be described in a later chapter.

CHAPTER 3 # The Drilling Machine

NEXT to the lathe the most important single item of machine equipment is the drilling machine. Whilst it is of course, perfectly possible to use the lathe for a number of drilling operations, indeed sometimes more conveniently, for the most part these are more easily and expeditiously performed by means of a machine specifically made for the purpose.

In the past a number of drilling machines have been designed and made ready for use by amateurs and others interested in the production of light machine components. But for one reason or another many of these have ceased to be manufactured and so, for better or worse, are unobtainable.

Today the drilling machines most suitable to the requirements of the small workshop can really be grouped into three main classes. The first a light high speed machine having a maximum capacity of ¼ in. The second a medium speed machine of ⅜ in. or perhaps ½ in. capacity. Whilst the third class comprises machines also of interest industrially, with a multi-speed range and a maximum capacity of as much as 1 in.

Of the first class probably the most successful, because it was specially designed for the amateur, is the 'Model Engineer' drilling machine. This drill, designed by E. T. Westbury, who will need no introduction to those already interested in light engineering, is intended to be made in the home workshop. The components have all been designed so that they can be machined on a lathe of 3½ in. centre height, and none of the work presents any real difficulty. A set of detailed drawings is available and castings may be obtained.

Next in the range of drilling machines suitable for amateur use are those illustrated in *Fig. 2* and *Fig. 3*. Both machines are the design of Messrs. E. W. Cowell of Watford and are, in part, suitable for making in the amateur workshop. They may be bought as sets of castings with some of the heavier machining already carried out or they may be had as fully finished machines.

The lighter of the two machines has a capacity of ⅜ in. and a three-speed

Fig. 7. The Champion drilling machine

Fig. 2. Cowell ⅜ in. drill

range of 800 to 2,500 r.p.m. or a six-speed range of 220 to 2,500 r.p.m. The dimensions of the work table are 6×5½ in. while those of the base, which may be used to secure work when necessary, are 9¼×5¼ in.

The maximum depth from the chuck to the base and the chuck to the worktable are 10½ in. and 8½ in. respectively, while the overall height of the machine is 24 in.

The ½ in. capacity machine, illustrated in *Fig. 3* is of heavier construction and, in addition to its duties as a drill press is also intended to be employed for light milling operations when used in conjunction with the makers compound slide bolted to the work table. To this end the whole quill assembly, that is the drill spindle and its bearings together with the rack-fed housing that supports them, is of altogether more substantial proportions than those of the lighter machine.

Moreover, the spindle is drilled for a drawbar needed to secure some forms of cutter and is bored No. 2 Morse Taper to accommodate the class of collet used with the Myford lathe. The nose of the spindle itself is threaded to accept the ring nut required to close these collets.

The machine has six speeds from 150 to 2,100 r.p.m. and its cone pulley floats on a separate pair of ball races thus relieving the spindle from all side thrust.

The third class of drilling machine is one extensively used in industry. It is of a design that appears to have had its origins in America, many machines with a family likeness being imported during the last war.

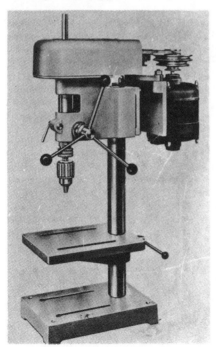

Fig. 3. Cowell ½ in. drill

The two drills we illustrate in *Figs. 4 and 5* come from the range made by Messrs. C. J. Meddings Ltd. of Slough. As readers will see the heads of both machines are identical only the mounting columns differ, one being intended for standing on the bench the other on the floor. Some may consider the floor model to be worth its extra cost, particularly when bench space is limited.

All the machines we have illustrated find a place in our own workshop so we are able to speak about them at first hand.

Driving the Drilling Machine

Whilst we have, in the main, described a class of machine that has its own built-in driving motor, there are still a number of drills not so provided. For these, and there must be not a few of them available on the second-hand market, a convenient driving method has to be contrived. Probably the simplest way is to mount the motors below the bench, allowing the driving belt to pass through the bench top in the manner depicted in the illustration *Fig. 6*. In this way the motor is well protected and occupies no valuable bench space. This method of mounting the driving motor has been used by the authors when fitting up the 'Model Engineer' drill and the earlier models of the Cowell drill.

The belts used with both these machines are of the round leather type $\frac{3}{16}$ in. dia. for the 'Model Engineer' drill and $\frac{1}{4}$ in. dia. for the Cowell machine, a subject that will be dealt with fully in a later chapter.

As drilling machines are only used intermittently they may conveniently be driven by low-voltage motors, but these need to be capable of developing $\frac{1}{4}$ or $\frac{1}{3}$ horse-power in order to function satisfactorily. This is a matter already

Fig. 4. Pacera bench drill

Fig. 5. Pacera floor drill

THE DRILLING MACHINE

Fig. 6. Mounting motor below the bench

Fig. 7. Speed range device for 'Champion' drill

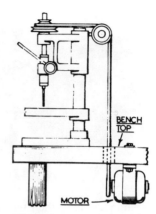

referred to in an earlier chapter where an illustration of such an application was given.

Speed Changing Arrangements

The simplest method of providing a range of speeds for the drilling machines suited to the requirements of the amateur shop is to fit the drill spindle and the driving motor with matching cone pulleys having three or more steps enabling the driving belt to be moved from one step to another without the necessity of adjusting its tension. This was the arrangement generally adopted in connection with most of the small drilling machines sold for amateur use.

When the driving motor was located **under** the bench top, however, there was naturally difficulty and delay when belt changing, so some form of speed selection device that could be located **on** the drilling machine itself was considered. The first of these fitments is that shown in *Fig. 7* where a 'Champion' ¼ in. machine, now no longer obtainable, is illustrated.

The system, apart from the jockey pulleys needed to allow the round leather driving belt to come up from under the bench, consists of a small counter shaft 'A' with three-step cone pulley mounted on an outrigger behind the head casting, an intermediate cone pulley 'B' attached to an adjustable arm secured to the top of the column and finally the original cone pulley 'C' fitted to the machine spindle itself. The belts used are miniature V-ropes that are readily obtainable, and these, after the speed selection has

Fig. 8. Speed range device for Cowell drill

been made, have their tension adjusted by the intermediate pulley secured to the column.

Virtually the same belt arrangement was fitted to the early Cowell $\tfrac{3}{8}$ in. Drilling Machine illustrated in *Fig. 8*. In this instance, however, the intermediate pulley is carried on a bracket secured to the head casting, while the countershaft is mounted directly on the machine column.

The more advanced and larger drilling machines to be had commercially usually have a wide range of available speeds attainable from multi-step cone pulleys. In addition, as the the Pacera machine depicted earlier, they are often fitted with a back-gear similar to a lathe, so the final range of available speeds is, therefore, doubled and much work such as tapping, the trepanning of large holes and operations of a like nature can be undertaken successfully. In addition many of the larger machines have electrical reversing switches fitted to them, a provision that greatly assists when simple tapping operations are being performed.

Choice of Chucks and Methods of Fitting

There is little doubt these days, that for drilling purposes the most popular and widely used chuck is that made by the Jacobs Companies. Years of experience both in design and manufacture have resulted in the production of high-quality components that, when assembled, provide a piece of equipment having great accuracy without which successful drilling to close tolerances is not possible. Before the general adoption of the Jacobs type of chuck there were numbers of different makes to be had, none of them renowned either for good design or quality.

For the most part, drilling machines have their chucks mounted on pegs or arbors. These have tapered seats to accept the chuck itself while the arbor may also have a shank machined to one of the Standard Morse Tapers enabling it to fit spindles similarly bored out.

Small drilling machines such as the 'Model Engineer' or the 'Cowell' however, have their spindle noses formed to accept the chuck directly as there is obviously no room in the spindle for an acceptance bore as large as the smallest standard Morse Taper. Both methods are illustrated in *Fig. 9* at 'A' and 'B' respectively.

In order to secure the chuck to the taper we have always advocated the

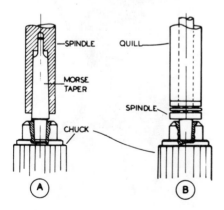

Fig. 9. Two methods of mounting the drill chuck

THE DRILLING MACHINE

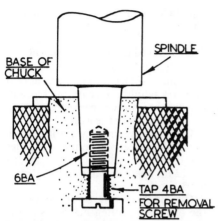

Fig. 10. Securing the drill chuck

introduction when possible of a screw set axially through the body of the chuck and engaging the drill spindle itself. A typical application of the idea is depicted diagrammatically in *Fig. 10*. It will be observed that the clearance hole for the axial screw is threaded so that a 4-B.A. jacking screw can be used to remove the chuck from the spindle if this becomes necessary. This is a method that is preferable to that of endeavouring to strike the chuck from its seating with the attendant danger of damage.

Before drilling and tapping a chuck in the manner illustrated, readers are advised to make sure that the particular chuck they wish to treat in this way can withstand this without detriment. An example of the type that must not be drilled axially is the 'Albrecht' chuck also made by the Jacobs Manufacturing Company. This is a keyless chuck, as opposed to the usual Jacobs product that needs a key to tighten it. The mechanism of the 'Albrecht' chuck is such that it could be damaged were an attempt made to drill the chuck axially. In any case many of the parts are hardened so drilling is not likely to be successful. These tools are extremely well and accurately made and have a range of

Fig. 11. The 'Albrecht' chuck

12 sizes with capacities from 0 to $\frac{1}{16}$ in. for the smallest to $\frac{1}{8}$ to $\frac{5}{8}$ in. for the largest chuck. For those interested in the drilling of very small holes may like to know that the first chuck in the range is provided with an engraved index on its body. This can be set to the size of drill that is to be mounted, thus making sure that it is properly gripped. Those readers who have had experience in these matters will scarcely need reminding how easy it is to insert a small drill, only to find that, when the machine is started, the drill point runs eccentrically because the shank is not properly seated in the chuck. Moreover, when very small drills are concerned, failure to mount them correctly at the first attempt could result in their permanent damage. The Albrecht chuck in question is illustrated in *Fig. 11*. In the interests of safety the bodies of modern drill chucks are now made smooth. In the

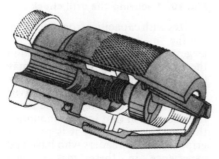

Fig. 11A. Section of the 'Albrecht' chuck

A High Speed Drilling Attachment

It often falls to the lot of the small workshop to carry out operations that industry would refuse on the score of expense or simply because there is just no equipment for the work.

One such operation was the drilling in stainless steel of a number of holes 0·007 in. dia., $\frac{1}{8}$ in. deep and spaced 0·012 in. apart in pairs on a platform 0·030 in. dia.

Clearly this was not work for any of the ordinary drilling machines in the author's workshop; but a little consideration showed that, if a suitable high-speed attachment could be devised, the old Champion drilling machine already referred to had sufficient feed sensitivity for success to be achieved.

The resulting equipment is illustrated in *Fig. 13*.

Here, an air-driven turbine 'A' is mounted in the drill chuck 'B' and is guided by the steady 'C'. The attachment, as has been said, is lowered to the work by means of the original feed handle and is counterbalanced, for the small amount of vertical movement it has to make, by the adjustable spring box 'D' forming part of the original drilling machine.

The steady is secured to the drill table and is adjustable so that a drill set in the chuck of the turbine can be brought to bear on the work. The drills themselves are caught in a special form of chuck called a collet. This device and other chucks are dealt with in Chapter 7.

The turbine is controlled by a needle valve and is fed from the shop air line

Turbines of the type employed were once largely used in connection with aircraft instruments. They are capable of very high speeds, up to 100,000 r.p.m. in some instances, so it is important that, when they are applied to

past several nasty accidents have occurred as a result of hair or clothing being caught by their serrated or knurled surfaces. In very small high-speed machines, where, for example, the diminutive Albrecht chuck would be used, this risk is negligible. But on large capacity drilling machines, where the drill mounting is usually at eye-level, a knurled chuck sleeve can be a hazard unless some form of protection is provided. Industrially this provision is now obligatory, but the amateur may well consider that to follow suit is a wise precaution. A typical chuck guard made for the Pacera machines is illustrated in *Fig. 12*.

Fig. 12. Chuck guard

THE DRILLING MACHINE

Fig. 13. High-speed attachment for drilling very small holes

drilling operations, these turbines have some means of control or the points of the drills they drive may be burnt.

Hand Drill Attachments

The makers of the more important electric hand drills are able to provide them with fixtures that will convert the drills into bench machines. While these conversions are, of course, capable of drilling metal with a fair degree of accuracy they seem to be best suited to woodworking, a field in which they are particularly useful and can be recommended.

Testing the Drilling Machine

While a new drilling machine of reputable make may be considered above reproach so far as accuracy is concerned, it may well be that after some time a second-hand machine, however well made, will have lost the accuracy it once had. Readers contemplating the purchase of a used machine may, therefore, consider it worthwhile to make a critical inspection of any drill offered for sale and, at the same time,

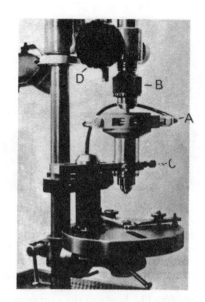

to carry out a couple of simple checks which will establish its general condition.

A quick glance will show if the equipment has been maltreated or not. Holes drilled in the worktable, and a general air of decrepitude will doubtless be a signal to investigate no further.

Fig. 14. Testing the drilling machine

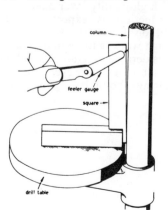

Fig. 15. Testing the drilling machine

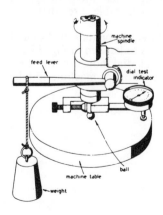

Fig. 15A. Testing the drilling machine

However, if the machine is clean and of good appearance, it will then be worthwhile to make the checks referred to. The first of these, illustrated in *Fig. 14* will give a rough estimate of the squareness of the work table in relation to the machine column. The blade of the square used must be of adequate length or an error may not be revealed. If the machine stands up to this check the would-be purchaser may proceed to the second test.

The object of this trial is to establish the accuracy of the work table alignment, in relation to the machine spindle itself; for it must be remembered that squareness with the machine column is not necessarily a guarantee that the drill spindle and the work table are themselves in alignment.

The equipment needed to make the check is to be seen in *Fig. 15* and consists of a dial indicator attached to an arbor set in the machine spindle. The outer end of the arbor is made to bear against a steel ball placed on the work table, and a weight hung on the work table ensuring that pressure is uniform. The dial indicator is set to zero at the position shown in the illustration, the drill spindle slowly rotated by hand and readings taken at four points around the drill table. If, at a point immediately to the front of the table, the dial indicator shows a reading of plus 0·001 in., this is in order, for it is usual to give a slight bias here in machines of quality so as to compensate for loads when actual drilling is in progress.

Unless the machine has a tilting table, not a highly desirable feature in the author's opinion, the readings at each side of it should show no discrepancy. If there is a variation the table will have to be adjusted, until there is no difference in the readings. When this has been done the checks at the back of the front of the table can be made, for these would be of no value unless the table had just been set level.

The amount of error that is tolerable will depend on the work the machine is going to be called on to do. If high-quality workmanship and production is expected, then the latitude is virtually nothing. If only rough work is to be undertaken then considerable tolerance is allowable. In all cases, however, the spindle bearings should be in good condition.

CHAPTER 4

Belt Drives

IN the past, machine tools in both the private and commercial workshop were driven by a single prime mover, such as a steam, gas or petrol engine, through the medium of lineshafting and individual countershafts, permitting the machines to be connected or disconnected from the drive as required.

Later, engines gave way to large electric motors, but the lineshafting and the countershafts remained, filling the roof area with a forest of flat belts that were an obstruction to the natural lighting of the workshop.

Today, however, each machine in the commercial shop and, in the main, most of those in amateur hands, has its own built-in electric motor. This has much simplified the driving of the tools for, as well as getting rid of expensive and light-obstructive lineshafts and belting, it allows any tool to be moved and re-sited without difficulty.

Accordingly, it is not proposed to devote much space to a form of drive that is now only of historic interest. Instead the emphasis will naturally be placed on current practice, and the conditions the reader will find at work both in the factory and in the private shop.

Nevertheless, since the use of flat belting cannot be dismissed out-of-hand, if only because there are still a number of the older machines driven by means of it, some brief notes on its use will appear later in the chapter.

Some years ago, in a booklet entitled *Belt Drives in the Small Workshop*, (Argus Books Ltd.), we went extensively into these matters. Those readers who possess copies of the book will no doubt pardon us if we can only deal with the subject somewhat repetitively in less detail here. However, they will find that information on such subjects as the driving of drilling machines or the provision of overhead belt systems for the lathe is included in those chapters that relate to the machine tools in question.

Round Belts

The round belt is certainly one of the most satisfactory means of driving small machine tools. It has great flexibility and can be used to transmit any drive which necessitates an abrupt change in angular direction since a round belt will run freely on jockey pulleys without any appreciable loss of power.

A good quality leather belt has much to commend it both on the score of expense and reliability. Today, however, round rubber canvas belts are being produced and these may be obtained in lengths that can be joined with a fastener or in the endless form if need be.

The fastener commonly used is a simple hook made from high-tensile steel wire. This, as will be observed in the illustration *Fig. 1*, is passed through holes drilled in the belting and bent over and clinched in the manner depicted.

However, if an irritating clicking noise is not to result as the belt runs over the pulleys, certain precautions must be taken when fitting the fastener as well as ensuring that the pulleys themselves are correctly formed.

'A' shows the form of the fastener commonly employed. If it is clinched

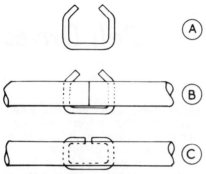

Fig. 1. Round belt fixing

over in the manner depicted at 'B', clearly the fastener will project, and become a source of noise. If, on the other hand, the fastener is sunk into the belting as demonstrated in the diagram 'C', then the noise will be much reduced if not actually eliminated.

It must not be supposed, however, that simply by squashing the fastener into the belt this end can be achieved; action on these lines would only result in disappointment. It is necessary, therefore, to cut shallow grooves lengthwise in the belt to allow the fastener to be sunk below the surface without destroying the belting itself. This is done by using a sharp chisel with the belt held in the vice and the cuts started from the drilled holes through which the fasteners are passed.

Fig. 2. Pulley sections

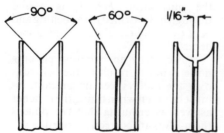

Pulleys for Round Belts

At one time pulleys used with round belting had a groove with an included angle of 90 degrees machined in them, as depicted in *Fig. 2*'. But experience has shown that, in order to transmit the maximum of power, the angle should be reduced to 60 degrees or less in order to materially increase the wedging action. In addition, as a further step towards silencing the drive, a rectangular groove is machined at the apex of the angle. This forms a clearance space for the fasteners should they protrude at all.

Jockey pulleys are commonly made with a circular groove. The reason for this is that, as no wedging action is needed here, in fact it must be avoided, the circular groove is best suited to allow the belt to run freely. Again, it is advisable to machine a clearance space for the belt fasteners.

Applications of the Round Belt

Present day uses for the round belt are principally confined to the driving of small drilling machines and grinding heads. Both aspects have been described fully in many other reference books so it is not proposed to further enlarge on this part of the subject.

However, a word should perhaps be said here on the subject of the lathe overhead drive, though this is a matter receiving attention in an appropriate and later chapter.

The lathe overhead is a belt system, driven either from the lathe power unit itself, or from a subsidiary motor, enabling work held stationary in the headstock chuck, or otherwise mounted on the mandrel, to be drilled or milled as required.

For the most part it is work needing only a little power, and this the round belt is well able to supply. In addition its flexibility enables the belting arrangements to be modified to suit

BELT DRIVES

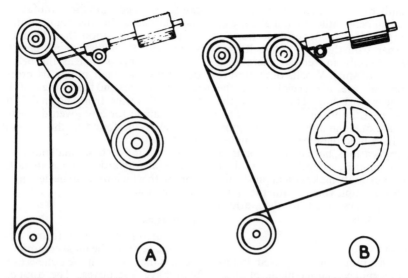

Fig. 3. Alternative belt layouts

almost any contingency. In *Fig. 3* two typical layouts are illustrated.

The layout 'A' is used when using a milling attachment mounted on the saddle of the lathe, whilst the arrangement 'B' was that employed for driving a spot-drilling attachment before it was fitted with self-contained electric drive.

The parts of the overhead drive are all adjustable positionally. Thus, if it is necessary to bring the jockey pulley forward over the lathe the counterweight can be moved back to compensate for this.

Reversing the direction of the drive when using a round belt is a simple operation. All that is needed is to cross the belt in the manner shown in the second diagram.

When the pulleys are well-designed and correctly machined round belts have a sound gripping action. It is, therefore, quite unnecessary to apply a heavy counterweight to the system. A weight of, say, 5 to 6 lb. will usually be quite sufficient to secure a satisfactory drive.

Round belting can usually be had in a variety of sizes, but for the most part, a $\frac{3}{16}$ in. dia., belt will be suitable for the small drilling machine whilst $\frac{1}{4}$ in. dia. is a convenient size to use in connection with the lathe overhead drive.

V-belt Drives

The popularity of the V-belt derives, for the most part, from the fact that it can be used to transmit power at very short centres. Typical examples are those seen in the illustrations *Figs. 4 and 5*, depicting the primary and countershaft drives of the Myford lathe. In both instances it will be seen that the centre distances between the motor and the countershaft, and the countershaft and the lathe mandrel itself, are commendably short, resulting in a most compact layout.

The smaller sizes of belt commonly used will transmit power up to a maximum of $\frac{3}{4}$ h.p.; they are very flexible and so do not heat up when bent round pulleys of small diameter.

For this reason, when increased power has to be transmitted, it is usually better to increase the number of small belts rather than make use of a single belt of greater cross section.

Apart from the range of V-belts that are generally in service there is a number of small V-belts of miniature form that are of especial use in the amateur workshop. They are, for the most part, approximately ¼ in. wide and are made in a variety of lengths. We have used these belts in a number of applications, both as single belts or in a composite assembly, where their great flexibility makes them most suitable for employment with the quite small pulleys fitted to the various mechanisms.

Pulleys

For the range of belts used industrially, pulleys, either made of sheet steel or cast light alloy, are available from stockists, many of whom are well qualified to give advice or solve any problem that may arise. The miniature belts referred to earlier, however, for the most part need pulleys that the amateur must make for himself.

In order to secure the maximum wedging action from a V-belt the groove in the pulley must be machined to the correct angle—information that can be obtained from the manufacturers catalogues. In this connection it should be understood that the angle varies with the pulley diameter. The reason for this is that as the belt is bent round a small pulley its cross-section changes and the V-angle of the belt itself varies from normal. The correct angle for any given diameter can usually be found in the relevant makers list.

The making of these small pulleys is a straight-forward exercise in lathe work with which we shall be dealing in a later chapter.

Applications

V-belt drives are in two forms. The first, and perhaps the most common of them, is that in which the pulleys used both have V-grooves. The applications of this drive can be seen throughout industry. The second form combines a V-pulley, the driver, with a normal flat faced pulley driven by two or more narrow V-belts. This type of drive is particularly useful because it often allows existing flat pulleys to be used. Examples in our own workshops are compressor drives and, in one instance, the primary drive of the lathe countershaft.

Fig. 4. Short centre belt drives

BELT DRIVES

Flat Belts
As has been said earlier, the flat belt has now fallen into disuse in the workshop except in a few cases where old type machines are in use. Nevertheless some notes on the use of the flat belt seem still justified.

At one time flat belting was made from a variety of materials such as canvas, rubberised canvas and leather. For the driving of machine tools, however, leather belting is still the most satisfactory since it is the least affected by oil and, for the most part, is the most flexible. It is probably the most expensive, but its all-round advantages and long-life, for drives 20 years old are not uncommon in the amateur workshop, make it, in the long run, still the best material available.

Belt Fasteners
Unless the belts are to be run over large pulleys the commercial pattern of fastener is really not suitable. The aim, in the private shop at all events, is a drive that is free from noise and is to all intents and purposes endless.

It is best therefore, to make a butt joint and to sew the two ends of the belt together in the manner seen in the diagram *Fig. 6*.

The sewing is carried out with 24 g. copper or soft iron wire after drilling the belt and cutting shallow nicks in its

Fig. 5. Short centre belt drives

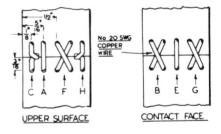

Fig. 6. Wire fastenings for flat belts

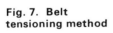

Fig. 7. Belt tensioning method

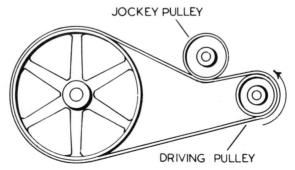

surface to allow the sewing to sink below the face in contact with the pulleys.

The number of sewings will of course depend on the width of the belt itself, but the diagram demonstrates an application to one having a width of 1 in.

Belt Tensioning Devices

Unlike the V-belt, if the flat belt is to be used with pulleys mounted at close centres some form of tensioning device will be needed. In the case of the lathe this may well take the form of a countershaft mounted on rocking centres so that the drive from the countershaft, and perhaps also from the motor, can be put under tension.

Perhaps the simplest course, however, is to provide a jockey pulley that can be brought into contact with the outside of the belt applying tension and increasing the wrap-round effect on the pulleys themselves.

In order to minimise any power loss the pulley must be mounted in contact with the free side of the belt and as close to the driving pulley as possible. This arrangement is depicted diagrammatically in *Fig. 7*.

CHAPTER 5 # The Shaping Machine

THE shaper, as it is commonly termed, whether it be operated by hand or power-driven, is a most useful machine for forming accurate flat surfaces and cutting long keyways as well as work of a similar kind. Although these operations can be carried out in the lathe by milling or fly-cutting, the shaper has the advantage, by virtue of its longer slides, that its capacity is greater for dealing with larger work, such as surfacing machine tables. Again, the method of mounting the cutting tool in the shaper head, and the massive castings used used for the sliding parts, make for a rigidity that may be lacking in a light lathe equipped with a revolving cutter.

A suitable tool enables a heavy initial cut to be taken when machining iron castings, and a subsequent light cut gives a highly-finished accurate surface.

The Drummond hand shaper, used for many years in the workshop, proved to be an accurate machine and was found capable of machining the table of a vertical slide measuring 6 in. by 5 in. Moreover, when tested on the surface plate, the part was found to be truly flat and did not require a final hand-scraping operation. In addition, the dovetail slides on the back of the casting were machined without difficulty. Although this machine, made for bench mounting or fitted with a pedestal stand, is no longer in production, the Cowell hand shaper is of somewhat similar design and appears to be a worthy successor.

There are also smaller machines on the market which are in some demand. The bench-mounted Cowell shaping machine, shown in *Fig. 1*, is of accurate and robust construction and capable of machining work up to a maximum of 6 in. by 6 in. The tool slide has a feed of $2\frac{1}{4}$ in. and the maxi-

Fig. 1. The Cowell shaping machine

mum distance from the tool to the surface of the table is 5 in. The five rates of automatic feed of the traversing slide range from 0·0025 in. to 0·0125 in. and the tool slide swivels through 180 degrees. The base member of the toolholder can be set over to afford relief on the idle stroke when taking cuts at an angle or on vertical surfaces.

It should be noted that, to reduce the cost of installation, the machine can be supplied in the form of a set of components with all the heavy machining completed, together with all necessary parts and materials. Included also are a set of working drawings and full instructions for completing the machine which can be carried out, like other Cowell products, in a 3½ in. lathe. A machine vice, designed for use with the machine, can be supplied either as castings or in the finished form.

To save working time and unnecessary labour, the Drummond workshop machine was replaced some years ago by the Acorn Tools power-driven shaping machine, which has proved in every way satisfactory both with regard to its accuracy and freedom from wear. In fact, despite much of the heavy work undertaken, no adjustment has been needed, although provision is made for this by fitting the sliding members with peel-off shims.

This machine, shown in *Fig. 2*, is massively built to ensure rigidity and a capacity for taking heavy cuts. The sliding ram has a stroke adjustable from ½ in. to 7 in. and is fitted with a graduated, swivelling tool slide mounting a toolholder and clapper-box. The machine table has a height adjustment of 5 in. and a horizontal travel of 9⅜ in. which is operated either by hand, or automatically to provide feeds of from 0·005 in. to 0·025 in. in either direction.

To prevent tilting of the machine table under the cutting pressure, the box-form casting is supported by an adjustable, travelling jack-screw. Both the tool slide and the traversing table are fitted with index collars graduated in thousandths of an inch.

The ram is driven by a ½ horse-power electric motor at the rate of from 40 to 170 strokes per minute through variable gearing incorporating a V-belt drive. Ball and needle-roller bearings are fitted to carry the driving shafts, and ample means of lubrication are provided for the working parts.

A swivelling machine vice is secured

Fig. 2. The Acorn Tools shaping machine

Fig. 4. Tool for cutting internal keyway

THE SHAPING MACHINE

Fig. 3. Shaping machine surfacing tools

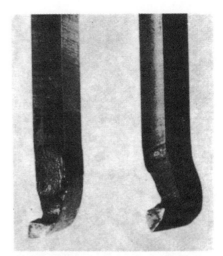

by means of T-bolts sliding in the slots of the box-form work table. The base of the vice is graduated in degrees to provide exact angular settings.

Operating the Shaping Machine

The usual high-speed steel lathe tools will serve for most machining operations.

The various tools to meet all ordinary requirements are fully described in the book *Shaping Machine and Lathe Tools*, published by Argus Books Ltd.

The tool shown on the right of *Fig. 3* is suitable for all general surfacing operations, and the second tool is designed for machining, at a single operation, work which has a flat, horizontal surface as well as an integral vertical shoulder.

Both these high-speed steel tools can be readily ground when resharpening becomes necessary. Tungsten carbide-tipped tools have been found to retain their sharpness when taking cuts through the scale of iron castings and, with a following light cut, they also give a high surface finish to this material.

The tool illustrated in *Fig. 4* was made for cutting internal keyways in the bores of pulleys and other machine parts.

The inset cutter-bit enables such keyways to be accurately machined for both width and depth by referring to the index collars fitted to the feed screws of the machine.

The Starrett taper gauge, *Fig. 5* is graduated in thousandths of an inch and provides a ready means of measuring the width and depth of the keyways as machining proceeds.

External keyways can also be cut in shafts with a tool similar to that used for parting off work in the lathe.

However, for this purpose, it is advisable to drill a shallow hole equal to the width of the keyway to form a run-out for the tool at the forward end of its travel.

After securing the work in the correct position on the machine table, and before starting the motor, the length of the ram stroke is set from the linear scale to enable the tool to clear the work at either end of its travel, and the position of the ram is adjusted accordingly.

It should be noted that the base of the clapper box needs to be set so that the tool is slightly inclined towards the work in the direction of feed. This enables the point of the tool to rise clear

Fig. 5. The Starrett taper gauge

with suitable rake and clearance angles, corresponding with those of the ordinary wood chisel or plane blade.

The Clapper Box

The tool slide of the shaping machine consists of two main components, the Base and the Clapper Box. As has already been shown the tool slide as a whole is attached to the head of the ram and can be turned at an angle in order to assist some machining operations.

It is sometimes necessary to lock the clapper box itself. Normally this is allowed to swing free in order to give relief to the tool, when machining overhung surfaces such as tool slides

of the work face on the idle or return stroke.

The rate of the ram travel will depend on its length of stroke; that is to say, time will be saved by using a rapid stroke for small work, and smoother working will be obtained with a slow stroke rate when machining large surfaces.

It is advisable to apply cutting fluid for machining steel, but cast iron is best machined dry.

When a batch of metal strips has to be machined parallel and to an equal width, this can be readily done by bundling or clamping them together in the machine vice, and then taking cuts over both the upper and lower surfaces.

Shaping operations on plastic materials give excellent results, and it has been found that Perspex, dealt with in this way, can be brought to a high surface finish by subsequently rubbing with a rag, backed by a wooden strip and charged with liquid metal polish. Woodwork can also be machined to a high finish in the shaper by using tools made from silver steel and formed

the box must be locked or the tool will jam on the return stroke.

Some machines, as they should be, are already provided with some means of locking the box. However, in an emergency when no provision is made, it may be locked by interposing a jack screw between the base of the tool slide and the back of the tool itself, as shown in *Fig. 6*.

Shaping Machine Toolholders

Two forms of toolholders are fitted to shaping machines, the English and the American. The first, comprises a cast-iron block fitted with setscrews projecting into a box-type seating to accept the tools themselves; the second, a pillar-like toolholder free to turn in a block forming part of the clapper box assembly itself.

The first has already been illustrated in *Fig.* 6, the second is depicted in *Fig.* 7.

From the standpoint of the shaping machine operator the American type toolpost has much to commend it, principally because it allows tools to

THE SHAPING MACHINE

Fig. 6. An improvised method of locking the clapper box

be set at an angle in relation to the tool-slide itself, and to a much greater extent than is possible with the English pattern tool holder. This facility is of great advantage when cutting down the side of work, an operation requiring the tool slide itself to be aligned at right angles to the surface of the machine table. When mounted in this way the variety of lathe tools that may be employed is much increased. It is no bad thing to collect together those tools that experience has shown to be particularly applicable to shaping processes and to mount them in a wood block so that they are readily available for use.

Additional Tools

Mention has already been made of the special tool for cutting internal keyways in the shaping machine. This piece of equipment is seen in action in *Fig. 8*. It is important to note that the tool must be **pulled** through the work, therefore the clapper box has to be locked so that the shaper can cut on the back stroke.

It is sometimes necessary to make an accurate saw cut in some particular component. For want of a milling machine this is often possible in the

Fig. 7. The American tool post

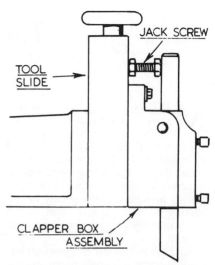

shaper using a short piece of hacksaw blade mounted in a holder that may be set in the tool post. It is obviously of importance that the saw is upright in relation to the work; this provision can, of course, be assured by testing set-up with a small square mounted on the machine table or the vice.

Fig. 8. Cutting an internal keyway

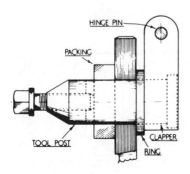

CHAPTER 6

The Milling Machine

MILLING as applied to workshop practice may be described as the process whereby rotating cutters of various forms are used to shape work by the removal of metal, their action being controlled to close limits of accuracy. Thus, by a milling process a keyway may be cut having both its width and depth dimensionally correct throughout its length.

So this book would not be complete without some reference to the milling machine itself. But it must be said at the outset that a really satisfactory and accurate machine is expensive and that the additional equipment needed increases the first cost considerably.

Furthermore, the provision of suitable cutters and equipment for keeping them sharp, an essential in milling operations, adds to the capital outlay.

For these reasons, then, milling machines are unlikely to be found in many purely amateur workshops though in the private shop, where it is possible to offset their cost and recover it by carrying out machining on a commercial basis, they may well have a place.

In the main, therefore, the amateur will wish to use his lathe for milling purposes, a subject that will be fully covered in Chapter 15.

In the past there has been a number of small bench machines produced for purely amateur use. It is difficult to say, however, whether a separate machine of necessarily somewhat light construction has any advantage in this respect over a robustly made lathe fitted with adequate attachments enabling at least the majority of plain milling work to be performed.

Our own experience both in the amateur and in the industrial field has taught us that, for the most part, milling in the lathe within capacity of the particular machine in use, is a sound practical proposition, and is capable of yielding results able to satisfy the most exacting requirements.

But to return to the milling machine. Many readers will know that, historically, the horizontal miller was

Fig. 1. The elements of the milling machine

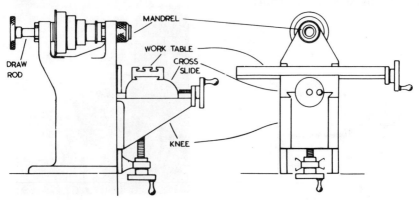

developed directly from the centre lathe. In fact early machines were little more than lathes provided with some means of feeding the work to the rotating cutter in a measurable manner. In addition means had to be found to support the cutter arbor against machining reaction; usually this problem was solved by some form of a tailstock similar to that used on the centre lathe.

By a process of evolution the horizontal milling machine has now become a unit comprising basically a main casting carrying a pair of bearings supporting the machine spindle, the latter provided with means of positioning and clamping the cutter.

The work to be machined is mounted on a table forming part of the knee assembly and is adjusted to the correct depth of cut by a feed screw set vertically in the knee. In addition, a horizontal feed screw controls the movement of the work table across the top surface of the knee. The elements of the machine are illustrated in *Fig. 1*.

A number of the small bench machines that are available have no other facilities than those of the basic machine. For many this will be no drawback particularly if the money saved by not providing automatic work feeds, back gear and other similar refinements is devoted to giving the machine the adequate spindle bearings and slides without which no satisfactory work can be performed.

With a milling machine components to be dealt with are either mounted directly on the work table or are gripped in a vice secured to it.

Machine Vices

The vices used on milling machines and other machine tools possess three main requirements. Firstly they must be accurate. Secondly they must be robust and lastly they must have low overall height. Whilst the first two requirements need no emphasis, a little consideration will show that much of the effective capacity of a small machine can be reduced by fitting to it a vice that is too high.

Reputable makers of machine tools manufacture the vices fitted to them. It can be assumed, therefore, that the equipment they supply is suitable and is also accurate.

Under these circumstances, therefore, it may not be considered necessary to check the vice supplied. On the other hand, if equipment is purchased from a source other than that of the machine tool maker himself, or has been bought at second-hand, then it is worthwhile applying some simple tests to it before putting the equipment into use.

In the case of the machine vice there are three principal points to watch for. These are depicted in the illustration *Fig. 2* where a vice having these major errors is shown.

In our illustration the standing jaw A, that is the one integral with the main casting, is out of square with the base C. The moving jaw B is loose on its slide and will ride up when it engages the work. Though not shown in the illustration the face of this jaw may not be truly parallel with that of the standing jaw. In passing it may be worth noting that one particular design of vice overcomes the last two difficulties by providing the moving jaw with some means of clamping it to the base once it has made contact with the work.

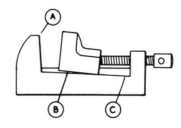

Fig. 2. Errors in a machine vice

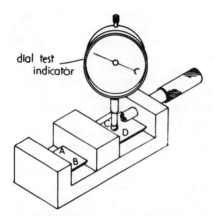

Fig. 3. Testing a machine vice

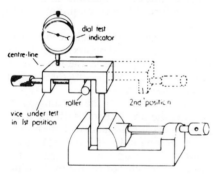

Fig. 4. Testing the standing jaw of a vice

Fig. 5. Testing the parallelism of vice jaws

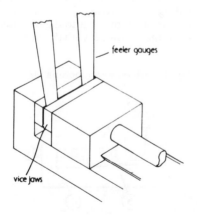

The last point to be observed is the parallelism of the work surface in relation to the base of the vice. This is indicated at (C) in the illustration, and by making a check this is the first consideration. The diagram *Fig. 3* shows how this is carried out. The vice itself is placed on a surface table and the contact foot of a dial test indicator is brought into contact with the work surface. The vice is then moved about under the indicator. If when the foot of the indicator making contact at A, B, C, and D, no movement of the indicator needle can be observed then both base and work surface are parallel. If accuracy is established we may proceed to test the squareness of the standing jaw. The method employed is illustrated in *Fig. 4*. Here a steel parallel of known accuracy is set in a vice also known to be accurate. The vice to be tested is then clamped to the parallel, a roller being interposed between the moving jaw and the steel parallel.

The dial test indicator is then applied in the way shown in the diagram. Parallelism in relation to the vice jaws themselves can easily be checked in the way depicted in *Fig. 5*.

Whilst the amateur can make good use of a simple milling machine provided only with manually operated work feeds, the small professional user will no doubt need more advanced equipment that has these feeds controlled automatically.

A typical machine is that illustrated in *Fig. 6* where one of the range of machines made by Messrs. Tom Senior is depicted. This tool is designed especially with the small user in mind. It has spindle speeds in twelve stages ranging from 60 to 4,000 r.p.m. enabling very small end mills to be used efficiently whilst the lower spindle speeds available allow large cutters to be employed for heavy stock removal.

The work table fitted has dimen-

THE MILLING MACHINE

sions 20 × 4¾ in. and a travel of 10 in. Its cross travel is 4½ in. The vertical travel of the knee supporting the table is 9 in.

As will be seen an overarm is fitted and along this the arbor support bearing can slide being locked in place after positional adjustment has been made.

The machine is in every way a thoroughly practical conception. The cutter spindle is carried in Timken tapered roller bearings whilst the front face of the main casting has a register tapped for screws holding the slotting attachment or other fitments.

A simple type of dividing head is available enabling most standard divisions to be obtained.

Perhaps not the least interesting point to the small user is that machined castings for the range of Senior Milling Machines can now be supplied.

The Vertical Milling Machine

Whilst a vertical milling attachment used in conjunction with a horizontal machine may suffice for much of the work to be undertaken there is no doubt that a miller designed for the purpose has many advantages over any such combination.

A good example of a vertical machine is that illustrated in *Fig. 7*. Here the milling unit with its driving motor is attached to an overarm thus allowing much latitude in the way the cutter approaches the work. In addition to the normal controls for the work table and knee the milling unit itself has its own independent feed arrangements and a depth stop that can be set to suit the work in hand.

Equipment

Whatever attachments are used with the milling machine there are certain additional items of equipment without which the tool cannot be employed to its fullest capacity.

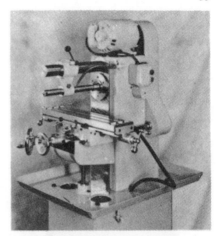

Fig. 6. The Tom Senior milling machine

The first of these is the Dividing Head. This comprises two units; 1. A small headstock with division plate, a mandrel with driver plate and centre, and a detent to engage holes or

Fig. 7. A vertical milling machine

Fig. 8. The dividing head

Fig. 9. The rotary table

Fig. 10. The rotary table

slots in the division plate itself. 2. A tailstock with centre adjustable in relation to the work which may be either mounted on centres between the two units, or carried in a chuck attached to the first unit.

In this way the part to be machined may be rotated through the successive fractions of a whole turn required to complete the work in hand. Both units are mounted on the work table of the milling machine as shown in *Fig. 8*.

The second item of additional equipment, is the rotating table. This is a device that is itself secured to the machine table enabling work placed on it to be turned through calculated parts of a turn with a high degree of accuracy.

Two forms of Rotary Table are illustrated in *Figs. 9 and 10* respectively where it will be seen that both have the edges of their work tables engraved in degrees and are provided with a vernier attached to the operating wheel enabling minutes of a degree to be indexed.

Both examples have tables T-slotted to accept fixing bolts and one has a pump centre allowing work to be mounted concentrically with greater facility.

Like the Dividing Head the mechanism of the Rotary Table comprises a worm and wormwheel, the latter forming part of the work table itself. It will be appreciated that no backlash can be permitted in the engagement of these gears. To this end some form of adjustment has to be provided ensuring that the worm and the wormwheel do mesh accurately. *Fig. 11* shows the wormwheel under the work table and the retaining plate attached to it.

Fig. 11. The rotary table wormwheel and retaining plate

CHAPTER 7

Chucks

BEFORE dealing with the different lathe operations two items of equipment need to be considered. The first of these are the various forms of chuck used to hold the work and the second, the tools used to do the actual machining.

Chucks

The chuck is one of the more important items of workshop equipment, and it is also one of the more expensive. Our experience, both in the professional as well as in the private sector of workshop activity, leaves us with the impression that chucks in all their forms deserve better treatment than they usually get.

Early lathes possessed no chucks as we know them. For the most part the lathes were used for wood turning and the work was either mounted between centres or fixed to an elementary faceplate. Subsequently work tended to be cemented into wooden cups or chucks and these in turn gave place to brass cups screwed to the nose of the lathe mandrel and furnished with a series of screws to secure the work. Such chucks are known as bell chucks and are sometimes found among the equipment of certain precision lathes though they have largely fallen into disuse.

The bell chuck seen in *Fig. 1* can, however, be said to be the forerunner of the 4-jaw independent chuck illustrated in *Fig. 2* an intermediate development being the faceplate fitted with dogs or jaws introduced by certain German manufacturers some 50 years ago.

The 4-Jaw Independent Chuck

In the 4-jaw independent chuck the jaws are reversible and so may be used to hold work by the bore as well as from the outside. Since the jaws move independently of each other the chuck may be used to hold irregular work

Fig. 1. The bell chuck

Fig. 2. The 4-jaw independent chuck

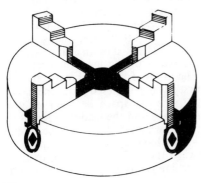

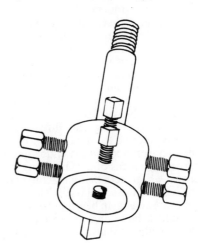

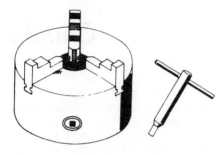

Fig. 3. The self-centring chuck

such as castings provided that these are within the chuck's capacity. The jaws have a portion of square thread on their back face engaging corresponding screws set in the body of the chuck itself. Under normal conditions the threads on the jaws are fully engaged. If, however, one tries to grip a piece of work too large for the chuck the threads are not fully engaged and those in contact may well be strained. Moreover, the bearing of the jaws in the ways machined in the chuck body is reduced allowing the jaws to tilt.

There is no difficulty when reversing the jaws in the body, for the latter is numbered to correspond with the mating jaws which should always be placed in their correct location.

The Self-centring Chuck

As with the Independent chuck the jaws of the self-centring chuck, illustrated in *Fig. 3*, are set in tenons formed in the chuck body itself. The jaws, usually three in number, carry threads on their reverse face corresponding with those machined on the scroll used to control the jaw movement. The scroll is a heavy disc of toughened steel incorporating the thread previously referred to; it has a bearing in the cast iron body of the chuck and is turned by means of bevel gearing, the crown wheel teeth being machined on the back of the scroll itself. The bevel gears are also set in bearings formed in the chuck body and are extended to the outside so that a T-handled key can be used to turn them when inserted into the internal square machined in the extension. Whilst it is possible to set work in an independent 4-jaw chuck with complete accuracy, this, for the most part, is not true of the self-centring chuck. For it must be borne in mind that the accuracy of this type of chuck is entirely dependent upon the skill and precision used in machining the scroll and, perhaps to a lesser extent, the jaws.

Changing Jaws in the Self-centring Chuck

The jaws of a self-centring chuck are not reversible so two sets have to be provided; one set called the drill jaws, are used for the holding of work by its outside surfaces as well as for gripping it internally on one or other of the set of steps formed on these jaws.

The second set of jaws provided with the chuck are employed when large diameter work, perhaps supported by the tailstock, is being turned. In this way the maximum contact between the scroll and the jaws is maintained. Both the body and the jaws are numbered so it is not difficult when changing jaws, to ensure that each is in its correct place.

The method of numbering is seen in the illustration *Fig. 4*. The jaws have to be replaced in the correct order or the chuck will not self-centre. Whilst the customary sequence is jaw No. 1, jaw No. 2 and finally jaw No. 3, this is not always so; therefore, the sequence must be checked by an examination of the backs of the jaws themselves as seen in *Fig. 5*. This will reveal that the first jaw to be entered is the one having the smallest space between its gripping face and the leading edge of the thread whilst the last jaw to be entered is that having the greatest space between them. The remaining or intermediate jaw is, obviously, entered be-

CHUCKS

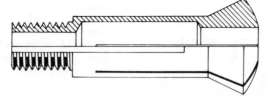

Fig. 6. The collet chuck

tween the first and last jaw. A look at the jaw numbers will now confirm whether the sequence is 1, 2, 3 or otherwise; and a note made for future use.

Since chucks tend to operate in an area where swarf and metal particles are present, it follows that some of this unwanted material may find its way into the scroll and on to the back of the chuck jaws. Before changing jaws, therefore, all swarf and metal dust must be removed either with a brush, or preferably with an airblast used intelligently, to ensure that both the scroll and jaws are clean before reassembly.

Both Independent and Self-centring chucks are fitted to the mandrel nose through the medium of a backplate. The plate is screwed to the mandrel and is machined with a register engaging the complementary recess at the back of the chuck. We shall be dealing further with this matter later when the fitting of chucks is discussed.

The Collet Chuck

In watchmakers lathes and most precision lathes a third form of chuck is commonly used. This is the collet chuck illustrated in *Fig. 6*. Essentially this chuck is a tube split three or more ways for part of its length and furnished with an angular nõse so that, when the chuck is drawn into the hollow mandrel of a lathe having a corresponding internal cone, it will contract and grip work placed within it.

Collet chucks are made to a high degree of precision enabling work to be removed and replaced with the certainty that it will run true. These chucks are particularly suitable for use in connection with instrument work of all kinds and watch and clock making in particular. In watchmakers and instrument lathes the closure of the chuck is effected by a draw-in spindle consisting of a hollow tube having an internal thread at one end engaging a corresponding thread on the collet, and a hand-wheel at the opposite end allowing the lathe worker to operate the collet. The arrangement is illustrated in *Fig. 7*.

For the newcomer three types of collet are of interest. These are illus-

Fig. 4. Numbering the chuck jaws

Fig. 5. Jaw sequence of entry

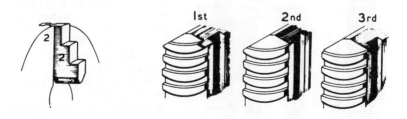

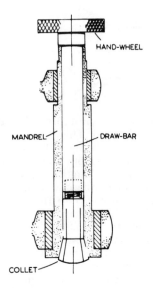

Fig. 7. Collet chuck assembly

and fitted to lathes of their manufacture. These collets fit into the tapered bore of the mandrel nose and are closed by means of a knurled cap engaging a ring machined on the outer extremity of the collet itself. *Fig. 9A and 9B.*

Since this ring is in fact an undercut, a simple piece of equipment has to be provided to close the collet and allow the cap to be slipped over before the collet is inserted in the lathe mandrel. This piece of equipment is illustrated in *Fig. 10.*

The Care of the Collet Chuck

The fact that collet chucks are pieces of precision equipment cannot be emphasised too often. They are costly and so, for the most part, owners of private workshops will give them fair treatment. Industrially, however, they seem, sometimes, to be the target for all sorts of abuse, so perhaps a few words on the care of collets and on their limitations may not be out of place.

First then, their limitations. Many years ago the then great authority on the precision lathe, the late George Adams, stated these limitations in most precise terms.

In the preface to what must now be thought an almost nostalgic catalogue of machine tools Adams says: 'In all the watch, clock and instrument trades split chucks are of the greatest importance. There is, in fact, no way of re-chucking work accurately other than by this method. The split chuck must be carried in a spindle with glass hard bearing surfaces ground and lapped to permit very high speeds while still preserving accuracy for long periods of time.

'A split chuck is also the only appliance with which one can re-chuck rods that must run true. There is really no latitude in a split chuck. It will only take just that size of true

trated in *Fig. 8.* The first of them marked 'A' in the illustration, is the collet usually found with precision or watchmakers lathes and is closed by a draw-in spindle.

The second marked 'B' is the pattern usually employed in connection with bar lathes and is closed by a cap acting on the nose of the collet.

The third form of collet, marked 'C', was introduced some years ago by the Myford Engineering Company

Fig. 8. Types of collet chuck

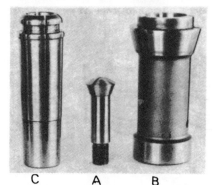

CHUCKS

Fig. 9B. Myford collet assembly

Fig. 9A. Myford collet assembly

cylindrical work which it is ground or lapped out to take.'

Obviously, then, if precision is the aim, one must not expect the collet or split chuck to accommodate oversize or undersize work and still run true. In fact the practice of distorting the chuck with unsuitable work may damage it permanently.

It will have been appreciated that the particular location of the split chuck in the nose of the mandrel renders it extremely liable to the ingress of swarf or metal dust. It follows, therefore, that when replacing work or changing the chuck itself both the chuck and its seating must first be carefully cleaned.

Keyless Chucks

Keyless chucks are principally used in the tailstock of the lathe to hold centre and other drills. An example of the keyless chuck is the Albrecht referred to in Chapter III. Albrecht chucks have great holding power and are obtainable in a wide range of capacities up to a maximum of $\frac{1}{2}$ in. diameter. The smallest of these chucks has a maximum capacity of $\frac{1}{16}$ in. and is provided with an index collar so that the chuck can be set to accept very small drills without difficulty.

Fig. 10. Myford collet closer and extractor

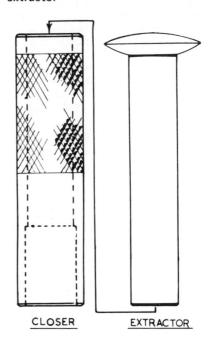

CLOSER   EXTRACTOR

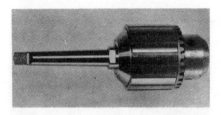

Fig. 11. The Jacobs chuck

Keyed Drill Chucks

The form of chuck most commonly used in both the lathe and the drilling machine is the keyed chuck illustrated in *Fig. 11*. The operation of these tools is similar to that of the keyless chuck, but greater force is imparted to the sleeve which closes the jaws by the employment of a key having a small bevel gear at its extremity; this gear engages teeth machined on the edge of the sleeve itself.

Mounting Drill Chucks

Drill Chucks are usually mounted on taper pegs fitting both the headstock and tailstock of the lathe as well as the internal taper in the spindle of many drilling machines. Some American lathes, however, have drill chucks screwed directly to the nose of the mandrel for use when machining small bar material, an application for which the drill chuck is particularly suitable.

It should be remembered that, in the case of lathes for the most part, it is only friction between the taper peg and the mandrel or tailstock, as the case may be, that prevents the drilling equipment from rotating under load. If rotation does take place both mating tapers may be damaged; it is important, therefore, to guard against this.

Where drilling machines are concerned the spindles are provided with a socket in the upper part of the female taper. The tang machined at the extremity of the chuck peg, and on taper shank drills, fits this socket and so effectively prevents the tapers from slipping. The same remarks also apply to some of the larger lathes. On most lathes, however, no such provision exists so the use of a draw bolt is recommended. A typical arrangement is seen in the illustration *Fig. 12*. The majority of drill chucks are secured to the peg itself solely by friction. If both chuck and peg are free from oil film and are really dry when put together the chances of slip are minimal. Nevertheless some may prefer to increase security by fitting a screw in the manner shown in Chapter 3 *Fig. 10*. An added refinement, if thought desirable, is the tapping of the axial

Fig. 12. Draw bolt fitting for chuck

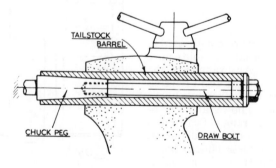

CHUCKS

hole in the chuck so that a much larger screw or bolt can be used to eject the peg if ever this becomes necessary.

A word of warning must be uttered, however, before this method of securing the drill chuck is adopted wholesale. Whilst, for the most part, chucks of the Jacobs pattern are so constructed that the scheme described may be adopted without any damage to the chuck, there are other forms that render the procedure impossible and any attempt to apply it damaging to the mechanism.

Fitting Lathe Chucks

Before we consider the steps that must be taken when fitting a chuck for oneself there are certain matters that must be fully understood.

Firstly, and for the most part, chucks are secured to the mandrel nose by a back-plate of cast-iron. This back-plate is secured to the nose by screwing, and it abuts against a shoulder formed on the mandrel itself. There is also a register on the nose of the mandrel. This serves to maintain the chuck back-plate in axial alignment with the mandrel thus relieving the threads of both mandrel and back-plate from the greater part of this duty. An examination of the illustration *Fig. 13* will help to clarify this point.

Secondly, there is a register machined on the face of the backplate itself. This register engages a corresponding recess formed in the chuck body. Because this recess and the register face at the back of the chuck are used as data during manufacture, it follows that the chuck itself will run true and concentric if the machining of the mating parts on the back-plate is accurately carried out.

The chuck is secured to the back-plate by screws, preferably of the socket type as these can be countersunk in the back-plate thus avoiding excrescences that might trap rag or lengths of swarf.

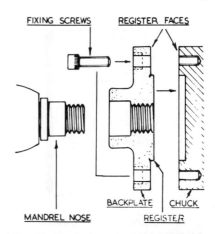

Fig. 13. Backplate and chuck assembly

The above remarks apply equally to independent 4-jaw or self-centring chucks, but in the case of the former of late years there has been a move to do away with the backplate and screw the chuck directly to the mandrel. Whether this move has received much encouragement it is difficult to say.

For the most part lathes are supplied with their independent and self-centring chucks ready fitted. Some manufacturers furnish an additional chuck back-plate ready screwed and correctly fitted to the mandrel nose ready for machining to fit a customers chuck.

The Care of Chucks

The first and perhaps more obvious way to ensure that a chuck will have a long working life and continue to give good service is to avoid putting into it work that is too large. This practice inevitably ruins the chuck in time because the gripping load, instead of being shared by a number of threads on both the scroll and jaws, is transferred to one or two threads with every likelihood of their partial or even complete collapse under the strain.

The second way is to avoid excessive tightening. The manufacturers of

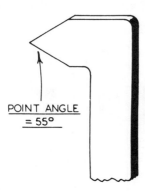

Fig. 14. Tool for removing swarf from backplate threads

chucks supply keys of the correct size for the purpose, so the practice of extending the length of the handle part of the key with pieces of metal tube in order to obtain greater leverage is to be deprecated. This deplorable measure may be observed only too often in industry; but it has nothing to recommend it and it will inevitably shorten the life of the chuck.

Finally, self-centring chucks need stripping down from time-to-time for cleaning and oiling. The reason for this is that, working as they do in an environment charged with metal dust, in time swarf may find its way into the inside of the chuck and set up wear to the scroll seating and other parts.

Some have advocated packing the chuck with grease on re-assembly. This seems a doubtful practice. Grease would seem to retain metal dust, whereas if thin lubricating oil is used small metal particles tend more easily to be thrown off by centrifugal force when the chuck is rotating. In lathes where the work is subjected to copious supplies of thin cutting oil we have always found that the chuck itself gets sufficiently well lubricated to need only infrequent attention.

When a chuck has been removed from the lathe it should be stood on the tips of its jaws with the back-plate upwards. This prevents any chips that may have found their way into the chuck from fouling the threads of the back-plate and so making difficult re-fitting to the lathe mandrel nose. But where this has happened the chips should be scraped away with a tool made from sheet brass in the form of a 'V' at one side, see *Fig. 14*.

When boring or drilling right through parts such as bushes or collars which are gripped in the self-centring chuck or 4-jaw chuck, it is important to prevent chips entering the interior of the chuck or the internal taper of the mandrel nose. This is best done by firmly plugging the throat of the chuck with rag or cotton wool. On withdrawing the plug, the accumulated chips will come away with it. Swarf entering the chuck can cause unnecessary wear and loss of accuracy as has been stressed earlier.

A final word in connection with independent 4-jaw chucks. It might be thought that by releasing two adjacent jaws of the chuck the work when remounted would again be centred by tightening these two jaws; but, with most chucks, an experiment, using a dial test indicator as a check, will show that this assumption is not realised.

The Chuck Brace

The changing of chuck jaws can be hastened appreciably if the brace illustrated in *Fig. 15* is used instead of the usual key supplied with the chuck. It will be appreciated, of course, that the brace is used only to bring the jaws into an approximately correct position in relation to the work; final adjustment is carried out with the chuck key itself.

The brace illustrated has the advantage that it can be used with either 4-jaw or self-centring chucks. As can

CHUCKS

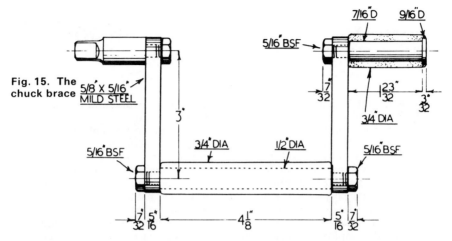

Fig. 15. The chuck brace

be seen from the illustration *Fig. 15* the brace is of built-up construction using bright mild steel sections of a class normally obtainable from material suppliers.

Something needs to be said about the general construction of the brace in case this should not be clear from the illustrations just mentioned. In particular some remarks need to be made in connection with the working end of the brace illustrated in *Fig. 15* and in detail in *Fig. 16*. In the first place the key for the 4-jaw chuck fits inside that for the self-centring chuck and is spring loaded so that it will project when required for use. It is locked by an allen grub screw in either the extended or retracted position. Secondly, in order to secure the large key against rotation in the lower member of the brace crank, a dowel is set axially in the key and is arranged to project into the brace member.

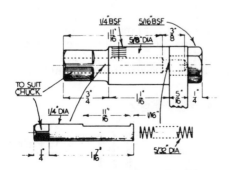

Fig. 16. The chuck brace

CHAPTER 8

Mandrels

MANDRELS are used to mount work for turning when it is essential that both the bore and the exterior of the work are machined concentric. A simple example is a bronze bush carrying a shaft. Concentricity, here, is of paramount importance especially when a pair of bushes have to be used in line.

Mandrels take many forms. That most usually employed is illustrated in *Fig. 1*.

A mandrel of this type is intended to be mounted between the lathe centres, and when correctly used is a highly accurate device. It consists of a case-hardened and ground steel bar, provided with female centres for mounting purposes, tapering very gradually from one end to the other. The extremities of the mandrel are reduced in diameter and have a flat surface milled on them to accommodate the carrier by means of which the mandrel is driven in the lathe. It is also usual to indicate the small end of the mandrel by a ring machined on it as seen in the illustration. A wide range of these mandrels is available commercially.

As will be appreciated work held in this way is secured by friction only. It is important, therefore, to make sure that any part being machined is as firm as possible and cannot slip. In the professional workshop this matter is taken care of by using a mandrel press, a piece of equipment, as its name suggests, specifically designed for the insertion and removal of mandrels.

The amateur is unlikely to have such a tool and so will have to rely on the copper hammer or raw hide mallet to ensure that the mandrel is well driven home. However, provided that the bench vice is large enough, it is possible, using suitable tubular packing and taking some care, to use it effectively as a substitute for the mandrel press.

Expanding Mandrels

In order to embody a wide range of expansion in a single device, expanding mandrels of the pattern illustrated in *Fig. 2* were introduced.

This, the Le Count expanding mandrel, consists of a body provided with female centres so that it can be used like a plain mandrel, and having three inclined keyways machined in it to accept stepped jaws upon which the work is set. The jaws can be moved along the keyways by an expander ring and so caused to grip the work.

Mandrels of this type have sufficient accuracy for many classes of work, but inevitably, as might be expected from the number of parts in-

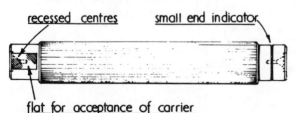

Fig. 1. Plain taper mandrel

volved, the overall accuracy cannot be as high as that of a plain mandrel.

Stub Mandrels

A common method of assuring concentricity in components is to machine them on a stub mandrel, or 'bung' to give it its colloquial name. These mandrels are made from short lengths of mild steel gripped, for preference, in the four-jaw independent chuck so that they can be machined accurately and reset by means of the dial indicator when it is necessary to use them again.

Usually the work is held by friction only so the mandrels take the form illustrated in *Fig. 3* at *A* and *D*. Nevertheless, there are times when it is expedient that the work should be secured more positively while at the same time being easily removed and replaced when repetitive machining is required. It is then that the mandrels seen at B and C are employed. The work is made a firm push fit on the device and is secured by a nut as seen in the illustration. These mandrels are useful, not only for machining work in the lathe, but also for handling work for milling purposes.

A somewhat novel device is that depicted in the illustration *Fig. 4*. This was produced in order to hold pressed-steel washers so that they can be machined to an acceptable standard. The washers are held by the bore and are forced against the face of the Body A by means of the threaded taper mandrel B. The parts comprising

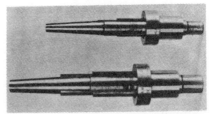

Fig. 2. The Le Count Expanding mandrel

Fig. 3. Stub mandrels

Fig. 4. Mandrels for holding washers

Fig. 5. Mandrels for holding washers

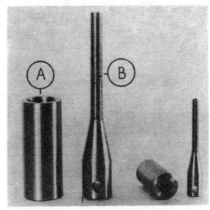

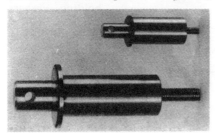

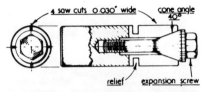

Fig. 7. Expanding stub mandrel

these devices are illustrated in *Fig. 5*. whilst their dimensional details are given in *Fig. 6*. Together the two mandrels cover washers from ⅛ in. to ½ in. internal diameter, a range that should suffice for all practical requirements in the small workshop.

The work that these devices are called on to perform is the holding of washers for the turning and chamfering of their edges, when only light cuts are needed. However, a centre may be drilled in the large end of the threaded member in order that the tailstock can be brought up in support if needed.

Expanding Stub Mandrels

In place of the plain stub mandrels referred to earlier it is sometimes convenient to use mandrels that have a limited range of expansion. These follow the basic design employed to mount a change wheel at the tail of the mandrel and described when dealing with methods of dividing in the lathe.

Expanding stub mandrels, therefore, follow the pattern depicted in *Fig. 7* and are easily made for oneself when needed. As with the plain stub mandrels previously described, if well made they can be readily set to run true by means of a dial indicator.

The amount of expansion available is small, about 0·005 in. in fact, and this movement must be assisted by turning a relief behind the working area as seen in the illustration.

Hollow Mandrels

There is often need to turn down square bar material in the lathe. Normally stock to be machined in this way would be caught in the 4-jaw chuck, and this still remains the correct method if a high degree of concentricity is wanted. However, an acceptable standard of accuracy can be

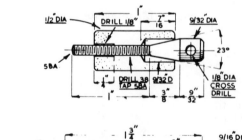

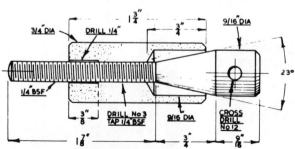

Fig. 6. Details of the mandrel for holding washers

Fig. 8. Hollow mandrels

achieved if the square material is passed through a hollow mandrel, the purist would perhaps call it a split collet, which is placed in the self-centring chuck and grips the material by its corners.

A group of these fitments is illustrated in *Fig. 8* one having a piece of square bar that has been turned down mounted in it. These mandrels are machined from mild steel bar, turned down to fit in the chuck and bored out so that the square material will slide into them without shake. They have integral stiffening collars at each end to preserve their concentricity and are split longitudinally in order that they will contract on the work when the chuck jaws are tightened. To remove any undue stiffness without, however, marring the accuracy of the fitment as a whole, the collars themselves are also split directly opposite the main saw cut.

The hollow mandrels shown were made to take $\frac{1}{4}$ in., $\frac{3}{8}$ in. and $\frac{1}{2}$ in. square material, sizes which are in common use in the workshop.

CHAPTER 9

Lathe Tools

AT one time the material for the making of the tools used in the lathe and other machines was almost exclusively Carbon Steel, a metal that could be formed, hardened, and tempered by the operative himself. But as the metals to be worked became harder and tougher, and the machining rates faster, the tool steels had to be improved. This requirement led to the production of high-speed steels of different specifications, very satisfactory from the machining stand point, but difficult for the operative to handle by himself when hardening had to be carried out. The heat treatment of carbon steel is a comparatively easy process needing the simplest of equipment that will be described in Chapter 25. Not so the hardening and tempering of the high-speed steels. These require more complicated apparatus including, for the most part, equipment that can measure very closely the temperatures in the heat treatment furnaces.

As a result more and more of the steel manufacturers tended to fully finish the materials supplied by them and so, today, we have available hardened and tempered tool bits that are ground all over. These need only grinding to shape before they are set in the lathe top slide. See *Fig. 1A*.

In addition some manufacturers can supply, in the finished condition, boring tools and form tools for screw cutting needing only sharpening when the occasion arises, matters that will receive attention in a later chapter.

These special tool bits are intended for mounting directly in the lathe under the top slide tool clamp or alternatively, in specially designed holders intended for a similar method of mounting.

There are many forms of lathe tool, and for these readers are referred to the book *Shaping Machine and Lathe Tools* published by Argus Books Ltd.

However, the essential tools are few and comprise those for turning, boring, screw cutting and parting off.

Front Tool

There are two forms of tool commonly used for turning the surface of cylindrical work. The first of these is the Front Roughing Tool illustrated in *Fig. 1*.

This tool is ground to the rake and clearance angles shown in the illustration, and the point is slightly rounded, the amount of rounding being increased if the tool is required for finishing purposes.

The direction of feed is from right to left as shown by the arrows, that is to say the tool cuts towards the headstock.

Fig. 1. The front roughing tool

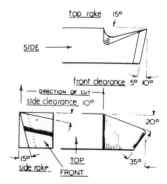

LATHE TOOLS

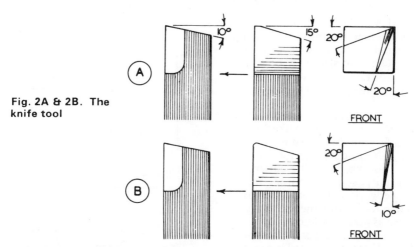

Fig. 2A & 2B. The knife tool

Knife Tool

The second and perhaps the most important tool is that depicted in the illustration *Fig. 2A and B*. This is the Knife Tool, more commonly used, for the most part, than any other. It may be employed both for sliding and surfacing and so is able to machine any parts that need shouldering as well as turning on their outer surfaces. In order to ensure that the finish imparted to the work is smooth the end of the tool has a slight land ground upon it. This tool is illustrated in *Fig. 2A* and can be used for both roughing and finishing cuts imparting a sharp corner to any shoulders. When these need to be rounded the tool is given the form shown in *Fig. 2B*.

The Boring Tool

There are several forms of boring tool, some are one-piece, others are boring bars having hardened tool bits mount-

Fig. 1A.

Fig. 4. The screwcutting tool

ed in them, the combination being gripped under the lathe tool clamp. The bar itself is sometimes set in a rectangular split sleeve, a convenient method of holding it since it enables the tool points position to be adjusted in relation to the work.

The clearance and rake angles of the boring tool are very much like those of the Knife Tool. In fact some workers prefer to reproduce the form of the Knife Tool when making cutters for boring purposes. The corner of the leading edge, in common with that of the knife tool, can either be sharp or rounded as occasion demands.

The clearance angles of a typical boring tool are as indicated in *Fig. 3*.

Screw Cutting Tools

The tools used for cutting screw threads in the lathe are of two types, one for the machining of male threads the other for the forming of their female counterpart. Both tools have the basic conformation depicted in *Fig. 4*.

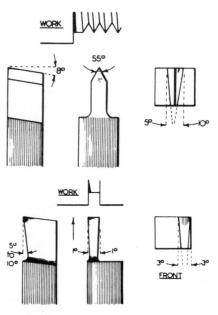

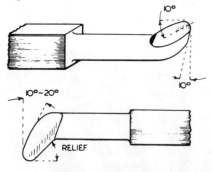

Fig. 5. The parting tool

The tool shown is one suitable for cutting male threads of Whitworth form. An internal threading tool is one similar to that used for boring but having the same point as that shown in the illustration.

In practice there are numbers of variants of the basic threading tool, all designed to simplify the production of accurate screw threads. These tools will be described in Chapter 21 where the subject of screw cutting in the lathe is dealt with in detail.

The Parting Tool

Once the piece of work has been fully

**Figs. 3 and 3B
Boring tools**

LATHE TOOLS

Fig. 6. The back toolpost

machined the operator will need to sever it from its bar of parent material. The process used to do this is known as 'parting off' and the tool employed is the Parting Tool depicted in *Fig. 5*. In order to lessen the stresses when this tool is in use, as well as to avoid chatter, the width of the cutting edge is made as narrow as possible consistent with adequate mechanical strength. Clearance angles are also reduced to a minimum to avoid weakening the support afforded to the cutting edge. Relief has also to be given to the tool behind the cutting edge to prevent jamming in the cut. The tool shown is suitable for parting steel and it will be noted that a top rake of some 5 to 10 degrees has been given behind the cutting edge. This rake should be extended well down the tool point, otherwise the coiled chips produced may form in the work groove itself and possibly jam, causing damage.

Fig. 7. The back toolpost and boring bar

The Back Toolpost

As a result of the forces that occur when parting off from the top slide set in its usual position in front of the work, difficulty is sometimes experienced in obtaining satisfactory results. This is the more noticeable when working with a light lathe, but it can be largely minimised if the tool is set behind the work. In this way the forces acting on the tool slide are upwards, thus thrusting its working surfaces into close contact and so promoting the greatest rigidity to the whole set-up. A little thought will show that when the tool is placed in front of the work the rigidity is impaired because the forces acting tend to separate the working surfaces, thus promoting instability.

Placing the parting tool at the rear of the work needs a special toolpost set at the tail end of the cross slide as seen in the illustration *Fig. 6*. Its holding down bolts are set directly under the toolpost thus providing the greatest rigidity. The device shown has provision for two tools, either being brought into play by rotating the capstan head and clamping it securely by the lever seen on the top of the tool-

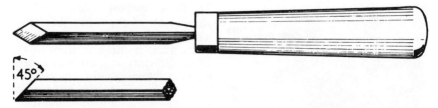

Fig. 8. The hand graver

post. The second tool may well be a chamfering cutter, as this is as often used as is the parting tool, leaving the top slide to carry any turning tool required.

The Back Toolpost is also useful to carry specialist tools such as boring bars and threading form cutters. An example of a back toolpost equipped with a boring bar is illustrated in Fig. 7.

When the Back Toolpost has a rotatable turret it is clearly important that, each time it is turned, the turret is correctly indexed and is always clamped in exactly the same place, whichever tool is selected. To ensure thus the turret is furnished with a dowel registering accurately in holes drilled and reamed in the body of the toolpost.

Hand Tools

At one time a great deal of turning was carried out with hand tools. Today, however, except for wood turning, the practice has almost ceased to exist and is confined to the rounding of corners, the turning of ball ends and similar work using the tool illustrated in Fig. 8.

The tool, which can be made from an old square file, or if greater durability is required, from a length of bright rectangular high-speed steel, has a lozenge-shaped facet ground upon it at an angle of 45 degrees to its shank. The latter is tapered so that it can be set firmly in a wooden handle leaving at least 4 in. of the shank projecting in order to provide a good hand hold.

Both hands are needed to use the tool which is set on a hand rest secured to the cross slide of the lathe and adjusted so that the cutting edges lie on the centre line of the work. In the case of a right-handed operative the left hand guides the tool whilst the right hand sweeps it in an arc using the left hand as a pivotal point.

Those readers who desire further information on this matter are advised to consult any of the lathework books where the subject is treated in some detail.

Tool Holders

At one time there were several makes of tool holder on the market, for the most part designed to accept the commercially produced tool bits already described.

Today the number available is much reduced and amongst those that can be purchased perhaps the most widely used are those made by Messrs. James Neill for their 'Eclipse' brand of tool bits. This company can supply a range of tool holders for turning, boring and parting off in sizes suitable for most lathes.

The amateur, however, can readily make such holders for himself and some examples designed to hold various forms of boring tool and made by us, are shown in *Figs. 9, 10 and 11.*

Tool Grinding

Unless the tools used are really sharp the quality of the work turned out can

LATHE TOOLS

only be poor. It is clear, then, that some form of grinder, however simple, is essential.

The subject of tool grinding is an extensive one so it is only possible to deal with it in an abridged form here. We have already covered tool grinding extensively in two books *Sharpening Small Tools and Shaping Machine and Lathe Tools* both published by Argus Books Ltd. In addition Argus has several other books with extensive coverage of the procedure.

The Angular Grinding Rest

Many readers will be aware that the sharpening process employed is called 'off-hand grinding'. The term is really self-explanatory for the method leaves to the operative himself the somewhat haphazard work of establishing the correct cutting angles on the tool freehand. There have been many devices described that will overcome this difficulty and perhaps the simplest of these is the grinding rest illustrated in *Fig. 12*.

The rest illustrated replaces those commonly found attached to electric grinders and others supplied commercially, and consists of but three parts; the Table 'A' supporting the tool during the grinding process, the Swing Arm 'B' that allows the table to be correctly aligned with the grinding wheel after being set to the desired angle and finally, the Angle Bracket 'C' securing the whole assembly to the bench of the machine. The adjustment

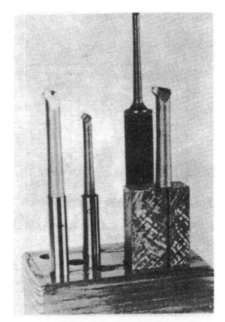

Fig. 11. Boring tool holders

Fig. 9. Boring tool holders

Fig. 10. Boring tool holders

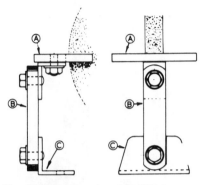

Fig. 12. The angular grinding rest

of the Table 'A' for a typical tool is illustrated in *Fig. 12A*.

Little need be said about the construction of an angular rest except to stress the necessity of providing a table of adequate size in order to give proper support to the tool being ground.

Fig. 13 illustrates a battery of angular grinding rests attached to machines of simple but very practical design once available on the market, whilst in *Fig. 14* a specially made grinder for finishing small tool bits is seen equipped with an angular rest of the type described.

The clearance angles required to be ground on the various forms of lathe tool are, for the most part few in number. So, in order to facilitate setting the angular rest and to avoid the unnecessary removal of tool steel

during the resharpening process, it is as well to provide oneself with templates that will enable the angular rest to be set accurately in relation to the wheels surface. The templates may take the form of a blade, say of $\frac{1}{16}$ in. thickness, either brass or steel will serve here, mounted in a footing, again either of brass or steel, enabling the whole to stand upright on the table of the rest and be brought into contact with the face of the grinding wheel while the angularity of the rest is adjusted. A typical template is depicted in *Fig. 15*.

Usually a pair of templates having angles of 5 and 10 and 15 and 20 degrees respectively will be sufficient and they should of course, as shown in the drawing, have their angularity stamped on them.

Grinding Wheels

Most tool merchants are able to advise on the correct grade of wheel to use for general grinding operations in the workshop but as a rough guide, whatever grade is suggested, the grit size should be 60 for roughing down and 80 for finishing the tool, possibly followed by a hand stoning operation with a fine slip stone in order to promote a keen, durable cutting edge.

The correct speed for a grinding wheel is one that will give it a peripheral speed of approximately 5,000 feet per minute. Thus a 6 in. wheel should run at 3,200 r.p.m. The following table gives the correct speeds for a selection

Fig. 12A. Setting the angular rest

SIDE OF TOOL   END OF TOOL   TOP SURFACE OF TOOL

LATHE TOOLS

of some commonly used wheel sizes:

Wheel diameter	Revs per minute
1	19,000
2	9,500
3	6,400
4	4,800
5	3,800
6	3,200

Truing the Grinding Wheel

The surface of the grinding wheel will need to be treated from time to time, both to restore the cutting ability of the abrasive grains as well as to ensure that the wheel as a whole is running true.

This is effected by a device called a wheel dresser, an example being illustrated in *Fig. 16* at 'A'. The dresser consists of a hand-held framework supporting a set of alternate plain and star wheels that are free to revolve when brought into contact with the grinding wheel. The action of the dresser is that of breaking away the dulled grains of the wheel and bringing to the surface new and sharp abrasive material.

The same result can be achieved by the dresser seen at 'B' this is an industrial diamond set in a holder so that it, too, can be brought to bear on the grinding wheel. While such a dresser, used in some circumstances, may have advantages, the first dresser described

Fig. 13. A battery of angular rests

Fig. 14. A special grinder with angular rests

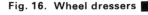

Fig. 16. Wheel dressers

Fig. 15. The grinding rest template

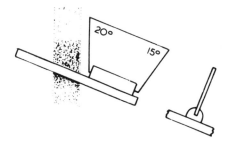

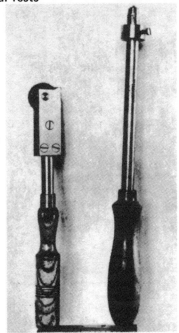

is that most commonly employed. It should perhaps be pointed out that much abrasive dust is produced by the wheel-truing process, so the location of the grinding machine should be as far away as possible from any m_ _hine tools. Otherwise the tools must be covered up whilst the process is being carried out.

A Rack and Stand for Lathe Tools

It is sometimes convenient to have those lathe tools commonly used, grouped together and close to hand. The stand illustrated in *Fig. 17* shows a rack for the tools together with a table to support small containers for cutting oil and such measuring equipment that may be in immediate use.

Fig. 17. Tool rack and stand

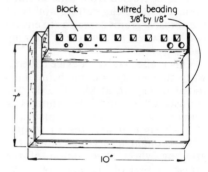

CHAPTER 10

Knurling

THE process of Knurling is employed to form a finger grip on cylindrical work. The knurling raises the surface by a rolling action, imparted by one or more serrated hardened wheels, and may take the form of a straight pattern with the serrations parallel to the axis of the knurl, or a diamond pattern produced by a pair of knurls having serrations set at an angular relation to their axis. The pitch of the serrations is the same for both wheels and they are set to cross on the work thus producing the diamond pattern. Some of the knurling wheels used in machines and the patterns they produce are illustrated diagrammatically in *Fig. 1*.

Formerly, when much work was produced in brass, often by hand-turning, any knurling needed had also to be imparted by hand methods. A single wheel, mounted in a suitable holder, was held on the hand rest and so positioned that a leverage was exerted upon the work. Obviously, the width of knurling that could be achieved was small being confined to such items as brass terminals, instrument parts and the like. The hand knurl wheel holder is depicted in *Fig. 2* whilst a selection of wheels suitable for mounting in it, and at one time available, are to be seen in *Fig. 3*.

As has been remarked, to be able to produce a reasonable knurled pattern even in soft materials it was essential to use considerable leverage. The diagram *Fig. 4*, illustrates the method of doing so. By this means only narrow surfaces can be treated because it is not possible to traverse the knurling tool along the hand rest at the same time maintaining even pressure upon the work.

Much amateur work is often spoilt by poor knurling and so, indeed, are many commercial products. So far as the amateur is concerned part of the trouble lies in the lightness of the

Fig. 1. Types of knurling

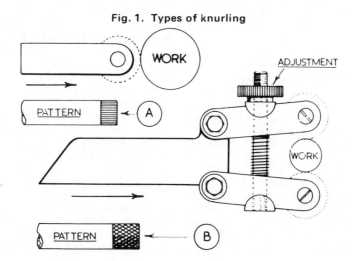

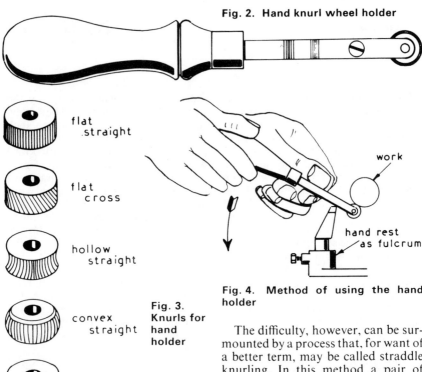

Fig. 2. Hand knurl wheel holder

Fig. 3. Knurls for hand holder

Fig. 4. Method of using the hand holder

lathes commonly in use. But, for the most part, the poor results stem from setting about the work in the wrong way.

The single wheel knurling tool illustrated in *Fig. 5* is unsuitable for use in a light lathe because of the damage it may do to the bearings of the headstock unless the tailstock can be brought up to support the work. The forces acting are directly at right angles to the axis of the lathe spindle, tending to force the mandrel out of line and do permanent damage. So, if a large component has to be knurled and a single knurl wheel must be employed, then the sequence of machining operations should be so arranged that the tailstock can be brought into play during the knurling process.

The difficulty, however, can be surmounted by a process that, for want of a better term, may be called straddle knurling. In this method a pair of knurl wheels are used, set in an adjustable holder, and are applied above and below the work in the manner depicted in the illustration *Fig. 6*. The wheels are adjusted to apply a slight wedging action, experience will show just how much this needs to be, so the resultant forces that act on the mandrel bearings will be no more, and possibly a good deal less, than those experienced during a normal turning operation.

The tool illustrated in *Fig. 7* and in *Fig. 7A* was made for use in a 4-in. centre lathe and has all the requirements for a holder designed for straddle knurling.

A similar device was designed and made by us to be used in the 3½ in. Drummond Lathe. Manufacturing drawings for this tool are given in *Figs. 8, 9 and 10* whilst the device itself is illustrated in *Figs. 11 and 12*.

KNURLING

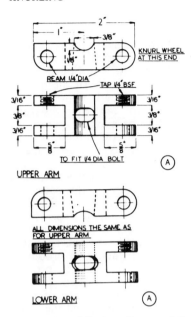

Fig. 9. Straddle knurling tool for 3½ Drummond lathe

Starting the Knurling Operation

In order to get the best results, and before the knurl wheels are set in motion, the work must run perfectly true. This may be ensured either by taking a light cut over its surface, or by setting it to run true with the aid of a dial indicator.

Fig. 8. Straddle knurling tool for 3½ Drummond lathe

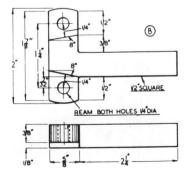

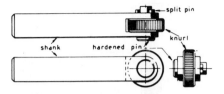

Fig. 5. Single wheel machine knurling tool

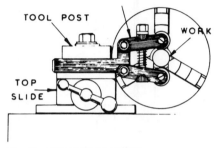

Fig. 6. Straddle knurling

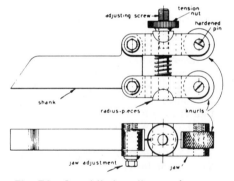

Fig. 7A. Straddle knurling tool

Fig. 7. Straddle knurling tool

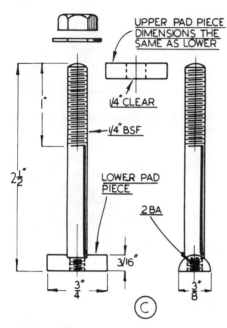

Fig. 10. Straddle knurling tool for 3½ Drummond lathe

When this has been done the knurls are engaged to straddle the work, making contact with at most ⅛ in. of the surface. The knurls are then tightened and the work turned by hand for a few revolutions.

Fig. 12. Straddle knurling tool for 3½ Drummond lathe

If, on examination, the pattern appears correct, tighten the knurl wheels further, engage the lathe back gear, also the automatic traverse, and proceed with the knurling. After a pass or two to establish the pattern throughout the length of the work the speed of the lathe can be increased and the knurling be allowed to proceed until the pattern, when viewed through a magnifying glass, appears sharp and its crests seen to be fully formed. However, it may be that when starting the knurling, the pattern is seen to be confused, the wheels being for the moment out of step. This is known as 'doubling', and it can be cured by increasing the wheel pressure until a proper pattern is produced. Thereafter the process can proceed following the lines indicated.

When a knurled surface of some length has to be produced the wheel holder must be set at a slight angle to the axis of the work. An angle of some 5 degrees is recommended, the purpose being to ensure a uniform pattern. This is particularly important when forming a straight knurl where the pattern must be perfectly parallel to the work axis. If the wheels are set square with the centre line of the work, there is then a tendency for metal to be thrown up in front of them, and they will not then follow a straight course but will be deflected; the resulting

KNURLING

Fig. 11. Straddle knurling tool for $3\frac{1}{2}$ Drummond lathe

pattern may then well be a slow helix composed of lines that are anything but straight.

Finally, a word about lubricating the work during the operation. It must be remembered that knurl wheels do not 'cut' in the strict sense of the word. Their action is one of crushing or squeezing, so metal is removed from the work in the form of fine particles. Unless a copious supply of suds or oil can be brought to bear on the work and wash this metal dust away as soon as it is produced, it is best where small areas are involved to carry out the operation dry. The application of small quantities of oil by brush or can does nothing to help and, in most instances, are definitely detrimental to the quality of the pattern produced.

CHAPTER 11

Lathe Operations

1. Mounting Work Between Centres

It is not the purpose of this book to go into the detail of elementary lathe practice. This has been fully covered in general lathe books. On the other hand there are certain processes or parts of them that need fuller coverage. The first of these is the matter of mounting work between centres. In order to do so the work itself has to be drilled so that it may accept the lathe centres as most readers will be aware. This drilling is performed with a special tool known as a centre-drill that provides a seating for the centre of the shape shown in the cross-section *Fig. 1* at A. The centre then engages this seating correctly as seen at B. From time to time, however, the centre drill needs resharpening, but all the operative himself can do in this respect is to grind the point of the pilot portion only since he has not the necessary equipment to sharpen the 60 degree cone portion as well. As a result, the relative positions of the two become progressively shorter and shorter till a condition is reached when, as depicted in *Fig. 1* at 'C', the lathe centre will not seat properly because its point is in contact with the bottom of the hole. We make no apology for calling the attention of readers to this defect; we have encountered it so often.

When drilling for centres it often happens that the work itself is too large to be passed into the hollow mandrel of the lathe. It must then be supported by the fixed steady in the manner illustrated in *Fig. 2*. Work mounted between centres is driven through a lathe dog bolted to the driver plate attached to the mandrel. In the absence of a forked lathe carrier, though these are readily made in the workshop, the standard form of carrier should be wired to the dog in the manner seen in the illustration *Fig. 3*. This will prevent the carrier knocking against the lathe dog during the turning operation.

2. Drilling from the Tailstock

When machining components in the lathe it is often necessary to use the tailstock to mount a drill. Tailstocks fitted as standard to the majority of lathes use a hand wheel in order to feed the drill to the work. Whilst the somewhat slow rate of advance resulting from this arrangement is, perhaps, tolerable when actual drilling is in progress it is certainly not so during the frequent withdrawals for clearing purposes that the drill itself requires; the more so if the drilled hole is deep. Moreover, the very fact that the tailstock mechanism functions on the

Fig. 1. The centre drill

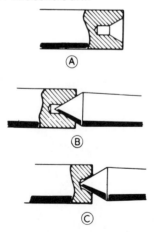

LATHE OPERATIONS

Fig. 2. Using the fixed steady to support work for drilling

Fig. 4. A lever-operated tailstock

The mechanism of the device depicted has the advantage of great simplicity. The lever anchorage is a clamp that may be swung round the tailstock casting allowing the operator to locate the operating lever in any convenient position.

An alternative system makes use of a rack and pinion to provide the feed. This arrangement is similar to that used in some drilling machines and is

Fig. 3. Wiring the carrier to the lathe dog

principle of the nut and bolt ensures that reaction forces, operating when the drill point is forced into the work, add a considerable friction loading that the operator must overcome. This is particularly in evidence when the drills are large.

All this can be surmounted, however, by making use of a lever feed, either as an attachment or, as illustrated in *Fig. 4*, by the substitution of a replacement tailstock so fitted. In this way the rate of in-feed will be greatly increased and the time lost in drill withdrawal largely reduced.

Fig. 5. The Cowell lever-operated tailstock for the Myford lathe

the mechanism Messrs. E. C. Cowell of Watford have fitted to their lever feed for the tailstock of the Myford Lathe, illustrated in *Fig. 5*.

3. Drilling Deep Holes

It may be worthwhile to point out to those readers who do not already know it that, when drilling a deep hole, one should not rely on the twist drill to keep to a straight path. Much may be done, however, to improve its performance by a modification to the drill point and this will be described in Chapter 18 'Drills and Drilling'.

Nevertheless, it is possible to produce an accurate and straight hole through the axis of the work by means of a tool called the D-bit. As its name implies the point of the tool is shaped like a letter D and this is applied to the work by means of an accurate

Fig. 6. The D-bit

chuck carried in the tailstock. The details of this tool are given in the illustration *Fig. 6*.

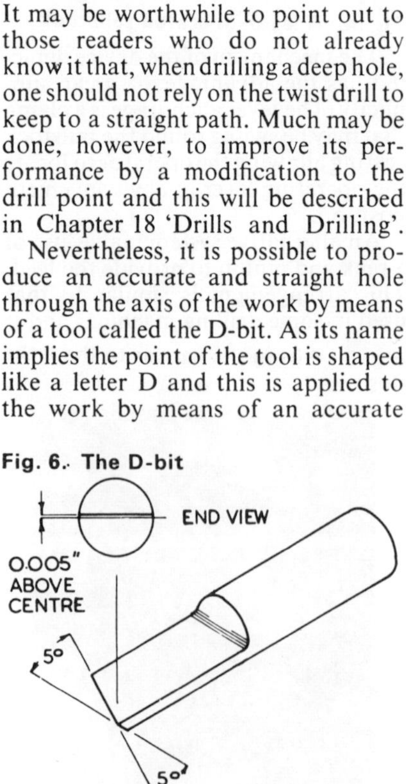

To use the D-bit the work is first drilled through axially some $\frac{1}{32}$ to $\frac{1}{16}$ in. less in diameter than finished size, and the hole is then opened out with a boring tool for a length equal to at least twice the diameter of the D-bit so that the latter will enter without shake.

In this way proper guidance will be given to the passage of the D-bit through the work when mounted in the tailstock chuck and fed in as seen in *Fig. 7*.

Plenty of lubricant should be used in the operation, both to promote free cutting as well as to oil the tools shank, in this connection many commercially made D-bits have provision for the introduction of cutting oil by the machining of a groove along their shanks.

4. Boring Work in the Lathe

The operation of boring in the lathe can be performed in two ways. The first is with the work gripped in the chuck and the boring tool mounted on the top slide. The second with the work secured to the cross-slide and a boring bar running between centres.

The tool used for the first method is shaped as shown in *Fig. 8* at A. This is the best form for reasons that will be presently explained.

We do not need to spend any time on the details of boring holes extending clear through the work. This has been done elsewhere. It is sometimes necessary, however, to bore holes that are blind. The initial procedure is first to centre drill the work, then to pilot drill it as shown in the illustration *Fig. 8B* to the full depth required. The hole is then opened out to size as shown by the dotted lines. Supposing, however, that the work is a bearing and that the shaft to fit it must make contact with the bottom of the hole,

LATHE OPERATIONS

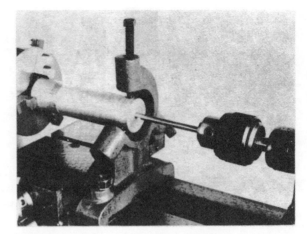

Fig. 7. Using the D-bit

then steps must be taken to clear the corner of the work or this requirement may not be fulfilled.

Removal of unwanted material is carried out by the tool depicted in *Fig. 8C* undercutting the work in the manner shown in *Fig. 8D*.

If one is boring a large hole the problem of making sure that the underside of the tool does not rub against the work is not a difficult one to solve, for the normal clearances given to the tool are usually sufficient.

When boring small holes, however, the position is very difficult. Much of the underside of the tool has to be ground away in the manner illustrated in *Fig. 9*. As one cannot possibly estimate beforehand how much the tool has to be modified, some way of doing so has to be found. The best

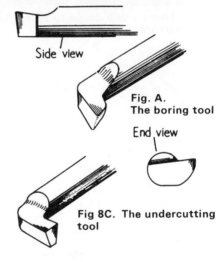

Fig. A. The boring tool

Fig 8C. The undercutting tool

Fig. 8B.

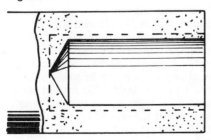

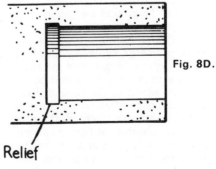

Fig. 8D.

Relief

Fig. 9A. Typical boring operation: gudgeon pin seating in an i.c. engine piston.

Fig. 9B. Counterbalancing work whilst boring

Fig. 9. Using a drill gauge to assess tool clearance

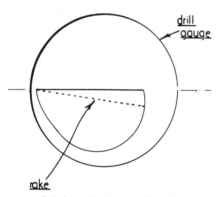

way of doing this is to refer the tool point to a drill gauge, selecting the correct hole size for the purpose. The amount to be ground away will then be apparent.

Two typical boring operations are illustrated in *Figs. 9A and 9B*. The first of these depicts the gudgeon pin seatings in a petrol engine piston being machined whilst in the second illustration the bore of an iron casting is having the same treatment. Both operations are examples of boring carried out on an angle plate attached to the lathe faceplate and illustrates the method used to counterbalance the work by bolting change wheels or other heavy metal objects to the faceplate.

It will also be seen that in the second illustration the top slide has been set over at an angle; this is to enable a fine feed to be given to the tool. If the top slide is set at an angle of 6 degrees then for every $0·001$ in. of top slide movement the toolpoint will receive an infeed of $0·0001$ in., a material provision when internal grinding operations in the lathe are being undertaken as will be seen later in the book.

It must be emphasised, however, that when applying the same procedure to a turning operation the tool used must be really sharp or no benefit whatever will accrue.

5. Boring Tool with Detachable Cutters

For some classes of work the tool illustrated in *Fig. 10* and *Fig. 11* has advantages particularly when shallow recesses have to be machined. The cutting tool itself has two seatings enabling a choice to be made suiting the work in hand. The first of these seats is at right angles to the device's axis whilst the second is at 45 degrees to it. The first position enables the tool

LATHE OPERATIONS

to be used for through-boring duties whilst the second location allows the tool to project when a blind hole has to be machined.

The details of the device will be clear from the illustrations where it will be seen that the hollow bar 'A' is provided with a cross-drilled thimble 'B' and that the tool bit passes through both these parts. The tool is secured by the push rod 'C' pressure being applied by the lock screw 'D'.

Making the equipment is not a difficult operation but it is absolutely necessary to make and use a simple

Fig. 10. Boring tools with detachable cutters

jig, the details of which are given in *Fig. 11A* in order to drill the cutter seating holes successfully.

Naturally, when boring a hole, the object of the procedure employed is to produce one that is both round and parallel. And in this the shape of the tool itself has a large bearing. If the illustration *Fig. 12* is examined the correct and incorrect tool shapes will be seen, and from this examination it

Fig. 11. Boring bars for small tools

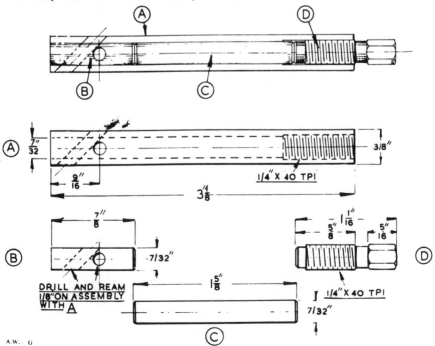

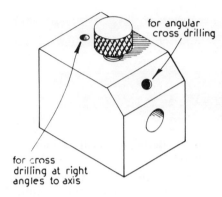

Fig. 11A. Jigs to ensure correct tool points

(labels on figure: for angular cross drilling; for cross drilling at right angles to axis)

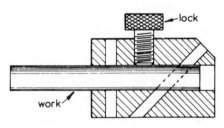

(labels: lock; work)

will be apparent that the tool point likely to give the best results is one that cuts on its front face only. The thrust from such a tool is directly parallel with the axis of the work, so the tool will not be deflected. Consequently the finished hole itself will

Fig. 12. Tool point shapes: right and wrong

be machined parallel. On the other hand, the thrust from a round nose tool is at an angle to the work's axis so the tendency is for a hole machined by it to become tapered.

When using a large and heavy boring tool this defect may not be in evidence, but in the size of equipment used when boring small holes it will almost certainly show itself.

The tool with detachable cutters is intended for mounting in a split sleeve clamped in the top slide. As such the amount of overhang, that is the distance between the tool point and the sleeve, can be adjusted to give the best results, and these are obtained, as depicted diagrammatically in *Fig. 13* when the overhang is reduced to a minimum.

A word should, perhaps, be said on the use of the split sleeve for clamping purposes. Unless the pressure of the top slide clamp is applied correctly there is every likelihood that the toolpoint will turn round in the work and ruin it. *Fig. 14* illustrates the right and wrong way of securing the tool.

Back Facing

Back Facing is a machining procedure sometimes necessary when it is important that both faces of a bored

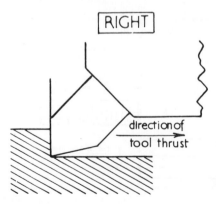
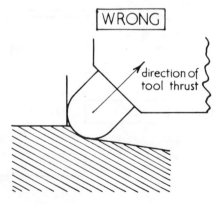

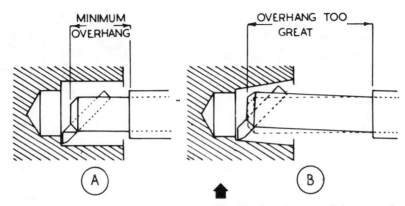

Fig. 13. Overhang—minimum and too great

hole should be exactly parallel one with the other; and when to remove the work from the chuck, then reverse and re-set it would destroy any possibility of accuracy resulting from such a process. *Fig. 15* depicts in diagrammatic form the type of tool used and the way it is applied to the work. The tool is similar to any other boring tool but has a somewhat longer head which is heavily raked so that the tool cuts on its back face only. In passing it is worth noting that the boring tool with detachable cutters described earlier, and for which detailed drawings were given, makes an excellent device for back facing. All that is needed is to reverse the cutter so that it points backwards and the equipment is complete.

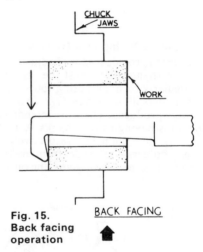

Fig. 15. Back facing operation

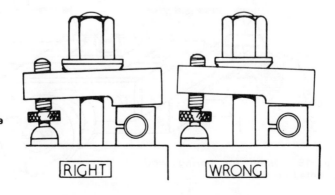

Fig. 14. Tool clamp pressure right and wrong

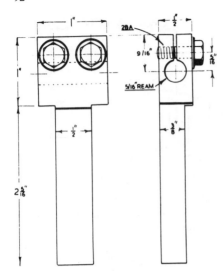
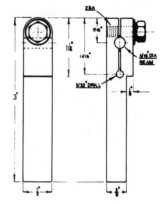

Fig. 17. Another simple boring tool holder

Fig. 16. Simple boring tool holder

Two Simple Boring Tool Holders

In order to provide means of adjusting the overhang that must be given particularly to small boring tools the two holders illustrated in *Figs. 16 and 17* were produced. These are intended for tools having round shanks $\frac{5}{16}$ in. and $\frac{3}{16}$ in. diameter respectively. These holders are made from mild steel to the dimensions given, the only comment needed on their construction being to make sure that the tool seatings are smooth and the holes accept the tools without shake.

Boring Work Mounted on the Saddle

The second method of boring work in the lathe is to mount it on the saddle. The elements of the necessary set-up are depicted in the illustration *Fig. 18*.

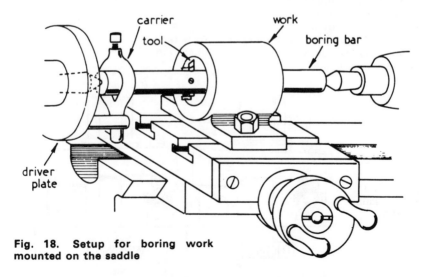

Fig. 18. Setup for boring work mounted on the saddle

LATHE OPERATIONS

This shows a typical casting bolted to the boring table of a small lathe where it should be noted, the bolts securing the work pass through it in such a way that their clamping effect on the T-slots results in their being supported by the face of the casting itself. If no support is given damage to the T-slots may easily result. The work is machined by a boring bar mounted between centres and driven by a carrier through a dog attached to the driver plate.

The Boring Bar

The Boring Bar, illustrated in *Fig. 19*, consists essentially of a substantial steel rod cross-drilled and reamed to accept the cutter, which is secured by a grub screw. In addition, however, some boring bars are fitted with a cheese-headed screw abutting against the end of the cutter enabling it to be adjusted the more readily and preventing it from withdrawing from the cut once it has been set.

The provision of a screw for this purpose is clearly only applicable to large boring bars where the added cross-drilling and tapping needed to provide a seating for such a screw would have no weakening effect on the bars themselves.

Setting the Boring Bar in Relation to the Work

If the work is to be bored correctly there are two essentials that must be established. First, the boring bar must be correctly centred in the work and then the point of the cutter itself must be adjusted to describe the correct arc. In order to satisfy the first requirement the work is marked off as seen in the illustration *Fig. 20* and placed on the boring table being roughly packed up to the correct height. The bar is then passed through and mounted between the lathe centres so that a 'Sticky Pin' attached to it can be applied to the work itself. The 'Sticky

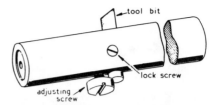

Fig. 19. Boring bar, with tool secured by grub screw

Fig. 20. Using the 'Sticky pin'

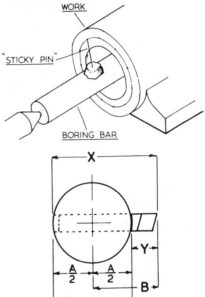

Fig. 21. Setting the cutter

Fig. 22. Measuring packing with dial test indication

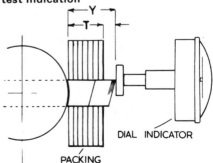

Fig. 23. Gearwheel being key-wayed

Pin' is nothing more than an ordinary pin used in sewing, set in a small lump of plasticene attached to the boring bar. If the lathe is now turned slowly by hand it will be possible to compare the path traced by the pin with the circle scribed on the work adjusting any packing until coincidence is achieved. Thereafter the work is secured and the cross-slide locked.

Setting the Cutter

Provided that the boring bar itself runs truly the cutter may be set with reference to the bar's periphery, and to do this there are two methods open to the operator of making the necessary measurement. In the first method the measurement 'X' illustrated in Fig. 21 is taken and from it is subtracted the dimensions $\frac{A}{2}$ that is half the diameter of the boring bar itself, thus leaving the measurement 'B' the radius of the arc described by the toolpoint.

The second method is to take the dimension 'Y' and add to it $\frac{A}{2}$ the figure thus obtained again being the radius of the tool point arc 'B'.

Making the measurement 'X' used in the first method is not always easy. If the cutter is seated in a blind hole then this can be obtained directly with a micrometer. But the conditions for doing so hardly ever obtain, so the second method must be adopted.

To make the measurement 'Y' a dial test indicator is used. If the range of the indicator is large, and the stand out of the tool point from the bar small, this dimension can be taken directly. To do so the foot of the indicator is applied to the boring bar and the dial set at zero. The foot is raised without moving the indicator itself and is then applied to the tool-

Fig. 24. Moving top-slide with a lever

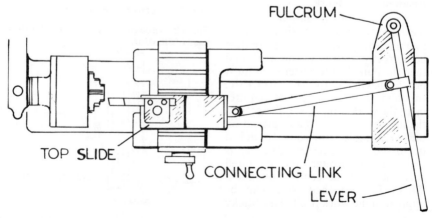

point care being taken to turn the lathe **backwards** by hand in order to obtain a maximum reading.

In cases where the tool projects too far for a direct reading to be taken, packing, measured with the aid of a micrometer, must be interposed between the boring bar and the indicator in order to set it to zero. The measurement 'Y' is then the thickness 't' of the packing, plus the reading obtained from the dial test indicator itself, as depicted in the illustration *Fig. 22*.

6. Cutting Internal Keyways

When making certain parts in the small workshop it is sometimes necessary to cut internal keyways. Though this is work that, in a large commercial undertaking, would normally be done in a vertical slotting machine, it is perfectly possible to accomplish the task successfully in the lathe. *Fig. 23* depicts a gear wheel being key-wayed. The cutter used is a parting tool mounted upon its side and packed to ensure that it is central with the work. To form the keyway the lathe saddle can be racked backwards and forwards whilst the tool is fed into the work taking cuts of about $0 \cdot 001$ in. to $0 \cdot 002$ in. at a time by means of the cross-slide. Using the saddle however, continuously in this way is not desirable. It is better to lock the saddle, thus avoiding local wear on the lathe bed, and move the top slide by means of a lever system clamped to the lathe bed itself, as illustrated in *Fig. 24*.

CHAPTER 12

Taper Turning

THE turning of tapers is one of the more important lathe processes undertaken in the workshop. The turner is often called upon to produce tapers, both male and female, that will match some existing mechanical part; and he must do so with complete accuracy or the fit of the mating components will be non-existant.

1. Turning Attachments

For those who have much of this work to do on a repetitive basis the expense of a turning attachment will be justified. The attachment seen in *Fig. 1*, usually bolts to the back of the lathe and consists of a slide that may be set over to the amount of taper required and is fitted with a cursor and a rigid connecting link securing the cursor to the cross-slide which, of course, has its feed screw removed for the period of the turning operation.

The slide can be set at an angle corresponding with half the included angle of the taper itself. If the saddle is now moved along the lathe bed the cursor will travel up the attachment thus moving the cross-slide in relation to the work and so reproducing the desired taper upon it. One of the advantages of employing an attachment is that the lathe self-act may be used to traverse the saddle itself. A disadvantage, however, lies in the fact that, with the feed screw disconnected the cut must be adjusted by slackening the connecting link and tapping the cross-slide forward. In many high class industrial lathes, on the other hand,

Fig. 1. Taper turning attachment

TAPER TURNING

matters are so arranged that the cross slide feed screw is kept operative and puts on the tool feed without impeding the action of the taper attachment.

2. Turning with the Tailstock set over

In default of a turning attachment long tapers can be machined by setting over the tailstock and supporting the work on centres during the turning operation. Indeed, this is the method most commonly employed, and all lathes of reputable make have provision for it.

In order to set the tailstock over it is made in two parts. A base 'A' and the Tailstock Body 'B' seen in the illustration *Fig. 2*.

A key 'C' fits accurately into both these parts so that when the nut 'D' is slackened, the tailstock body can be moved across the base without endangering the alignment of the tailstock's axis with that of the lathe itself. This movement is controlled, for the most part, by two screws set either side of the tailstock body and engaging a tenon projecting from the base. Adjusting the screws moves the body by amounts that may be read off on a scale set at the back of the base casting as seen in the illustration.

The degree of set over is usually limited to some 10 degrees each side of the central point.

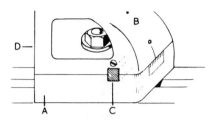

Fig. 2. Set-over tailstock

Fig. 3. Adjustable centre

3. Turning with a Set-over Tailstock Centre

Despite the comparative ease with which the tailstock as a whole can be set over, many operators are reluctant

Fig. 4. Adjustable centre dismantled

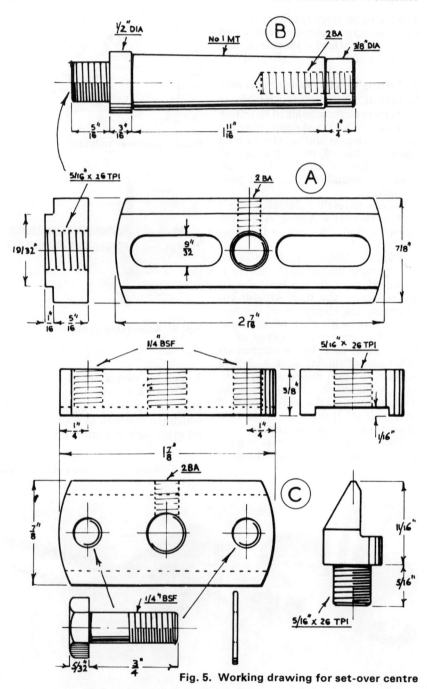

Fig. 5. Working drawing for set-over centre

to disturb it once it has been set to turn work perfectly parallel. However, the tailstock itself need not be set-over if an adjustable centre is employed. This piece of equipment is illustrated in *Fig. 3* and dismantled in *Fig. 4*. The device consists of two main parts, the Base 'A' and the Slide 'B'. The former is fitted with a Morse Taper peg having provision for a draw bolt so that the centre may hold securely in the tailstock. The slide, secured to the base by a pair of set-screws, is fitted with a screw-in half-centre. It is perhaps hardly necessary to point out that the slide must be set perfectly parallel with the face of the lathe bed or the object of the device will be set at nought. For this reason the parallelism should be checked with a dial indicator set on the cross-slide and applied to the Base 'A' itself. Working details of the set-over centre are given in the illustration *Fig. 5*.

4. Turning with the Top-slide Set Over

Much of the work the lathe operator is called on to perform consists in the machining of short tapers whose length is within the range of the top-slide's travel. For this reason the inclination required is most conveniently obtained by setting over the top-slide itself. Most lathes have the base of the slide engraved in degrees and with a zero or line incised on the face of the cross-slide. The top-slide may thus be set directly to the angle required. For a lathe not so equipped, or where there may be some doubt as to the accuracy of the engraving, the procedure depicted in the illustration *Figs. 6 and 6A* should be employed. Here, as will be seen, the top slide is removed from its base and a protractor, set to the angle required, has its blade applied to the base, a parallel test bar being interposed for the purpose. The base of the protractor engages the faceplate and the top slide is adjusted until coincidence with

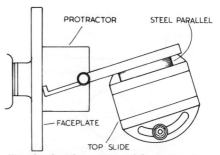

Fig. 6. Setting over with protractor

Fig. 6A. Using the protractor for an actual job

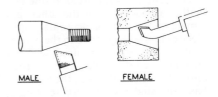

Fig. 7. Machining male and female cones

these two reference faces is achieved.

When set the top slide is used to machine male tapers, but the same setting can be used for the turning of corresponding female cones provided the turning tool is used inverted as shown in the diagram *Fig. 7*. A word of warning, however, must be uttered in relation to this and other turning operations; at all times the tool point must be on the centre line of the work or it will be impossible to machine an accurate taper.

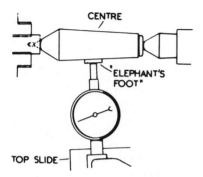

Fig. 8. Dial test indicator used to set taper.

5. Setting Tapers with the Dial Test Indicator

The Dial Test Indicator may also be used to set the top slide at the correct inclination for a given piece of work. A good example of the method employed is that illustrated in *Fig. 8*. Here we see a Morse Taper centre set up in the lathe so that the top slide may be adjusted to enable a replica of the taper to be machined accurately.

The centre is supported at one end by the tailstock and at the other in a female centre drilled accurately in a piece of material gripped in the chuck. A dial test indicator fitted with an 'elephants foot' or flat anvil is then applied to the centre and the top slide base adjusted until a zero reading is recorded by the indicator when the top slide itself is moved along the base.

Once the top slide has been set and firmly secured the check equipment is dismantled and replaced by the piece of material from which the part required will be turned. It is usually best to support the work by means of the tailstock and readers are again reminded that the toolpoint must be set on the centre line or no accuracy will result from the turning operation.

6. Calculating Tapers

The amount of taper on a given part may be expressed in two ways. In the first the amount of taper per foot is the standard used whilst in the second the cone angle of the part itself is stated. The first method applied, for the most part, to long slow tapers, whilst the second is used in connection with short components of a comparatively quick taper.

However, this may be in the final analysis the operator needs to know to what extent he must set over his tailstock or deflect the top slide in order to be able to machine the taper required. This involves a simple trigonometrical calculation the working out of which will provide the set-over angle he needs. In *Fig. 9* this matter is depicted in diagrammatic form.

In the diagram 'x' is the angle required. The value of this angle is $\frac{a}{b} = \tan x$ where $a = \frac{A-B}{2}$ and b is the

Fig. 9. Calculating tapers

Fig. 10. Calculating internal tapers

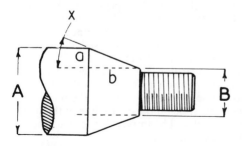

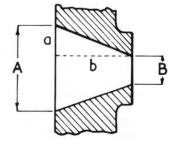

TAPER TURNING

Fig. 11. Checking tapers

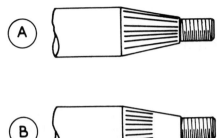

distance between the major and minor diameters of the taper. Both A and B are measurable with a micrometer whilst the distance b, preferably measured by means of a height gauge, can be assessed reasonably well with a depth gauge.

The assessment of an internal taper follows the same pattern but the dimension 'b' can be measured rather more accurately either with a micrometer or with a depth gauge.

The dimensions A and B seen in the illustration *Fig. 10* may be measured with a taper gauge when available or with internal callipers transferred to an outside micrometer.

7. Checking Tapers

It must not be supposed that, by using any of the methods described, a perfect mating of two tapers can be achieved without making a check during the progress of the work, and then carrying out any adjustments to the machining equipment that may be needed.

A simple method of making a check is illustrated in *Fig. 11*. Here at 'A' a male taper is depicted having its surface scored with lines applied with a soft lead pencil. If now the mating part is wrung on to the male member and then withdrawn the extent of the fit will be apparent from the amount of marking that is left on the taper. When over half the marking remains as seen in *Fig. 11B* it is usually advisable to make a machining adjustment. Slight discrepancies, however, may be corrected with a fine file whilst the work is rotating until, when the female taper is applied all trace of the marking is removed.

CHAPTER 13

Lapping

LAPPING is a process used to finish shafts and their bearings to close limits of accuracy, imparting at the same time a surface finish that will materially prolong the life of the bearing assembly.

The process is carried out by means of a lap, a piece of equipment charged with abrasive compound that rubs down the surface of the parts at the same time giving to them an extremely smooth finish.

The machined surface of parts before lapping consists, for the most part, of a series of small hillocks interspersed with valleys. Consequently, when, during use, the excrescences have worn off a pair of mating components (a shaft and its bearing for example), a good fit is often quickly reduced to an easy one having a low life factor. If, however, the wearing down process is carried out by lapping then it is possible to fit the parts very closely at the same time ensuring they are truly round.

Unless special compounds are employed not every material used in engineering is a suitable subject for the lapping process. Those metals that may be are comparatively few. Cast iron and steel, in both the hardened and soft condition, can be lapped, but the non-ferrous metals, and in particular white metal, must not be given the treatment because they become impregnated with the usual abrasives employed and are 'charged' as the term is. This makes them into good laps but renders them useless as bearings.

The Lapping of Shafts

When applied to steel shafts the lapping process serves to correct any minor errors of roundness or taper whilst, as has been indicated, imparting a working surface that will ensure a long life. It is usual to allow excess material, from 0.001 to 0.002 being usual, so that when fully lapped the finished shaft diameter is as laid down in the detail drawings.

The lap consists of a piece of cast-iron or copper, held in a holder as illustrated in *Fig. 1*.

The lap is split longitudinally, three ways, one saw cut passing through its wall. In this way the device can be contracted on to the work using the three screws provided with the lap holder. It should not grip the work too firmly, but should float axially along it not being allowed to dwell at any point for

Fig. 1. The simple lap and holder

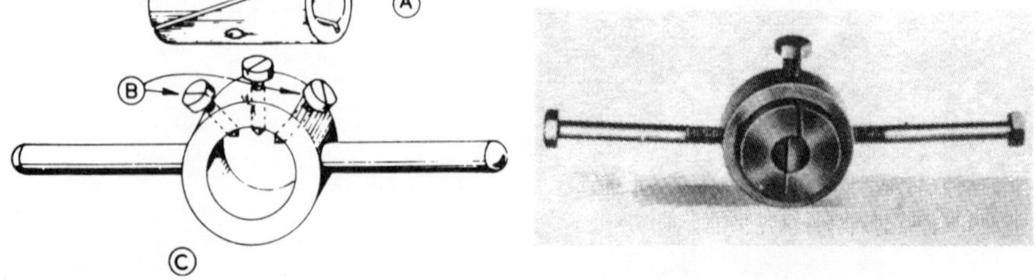

LAPPING

fear of 'ringing' the work surface. Abrasive compound is fed into the lap through the open sawcut, recharging taking place from time-to-time after the lap has been washed in petrol to remove the metal dust formed during the process.

Lapping is usually carried out in the lathe, the work being run at a medium speed, say 100 r.p.m. Naturally, the machine needs protection against possible ingress of abrasive compound, and this is best effected by covering the bed and any working parts likely to come into contact with the lapping compound. The coverings can be either of newspaper or rag and these, for obvious reasons, should all be burnt once the work has been completed.

A word of warning must be given about the use of a micrometer on a lapped surface. Make sure that the work is clean; for any abrasive left on it will in time, damage the anvil and spindle of the micrometer. The work, then, needs to be washed clean before measurements are taken.

Fig. 2. A simple expanding lap

Internal Lapping

The procedure used for lapping internal surfaces such as cylinder bores is similar to that used for external

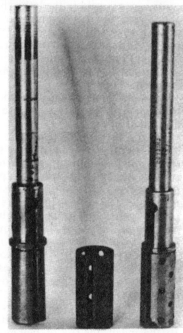

Fig. 3. The Boyar-Schulze lap

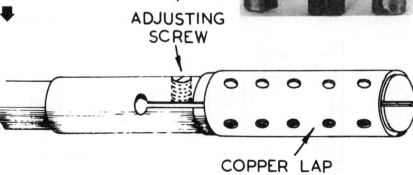

ADJUSTING SCREW

COPPER LAP

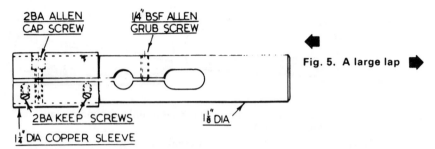

Fig. 5. A large lap

work. The tool used is an expanding lap similar to that depicted in the illustration *Fig. 2*. The shaft of the contrivance is held in the lathe chuck and has a tapered seating for the lapping head which is driven up the taper in order to expand it. Naturally, only light taps with a raw-hide mallet are needed to obtain any expansion required.

A somewhat more sensitive lapping instrument is the Boyar-Schulze lap illustrated in *Fig. 3*.

As with the adjustable lap just described, the Boyar-Schulze has a steel shank split longitudinally and furnished with an adjusting screw to expand the copper lap surrounding the head of the shank. The lap is made from sheet copper, perforated to hold the abrasive compound, and three copper heads are provided with each tool. The user is thus able to keep a separate head for each of three grades of abrasive, an essential condition for the finest work.

Internal lapping is commonly undertaken with work held in the hand. By this means the risk of 'bell mouthing' the part is, in the main, eliminated, because the work floats on the lap and is not subjected to constraint in any direction.

However, when the component is too large to be held in the hand it must be mounted in the chuck whilst the lap is supported by the tailstock. On no account must the lap be held rigidly but be carried flexibly in the manner depicted in the diagram *Fig. 4*.

The rubber hose used should be of the canvas-impregnated variety, sufficiently strong to resist the torque of the lap but flexible enough to impose no restraint on the lap. In this way the tool will follow the line of the work and not be pushed sideways, with a result that the bore may become bell-mouthed.

Copper laps of the type illustrated in *Fig. 3* are not difficult to make. The copper head is readily rolled up from sheet material after drilling the perforations and dressing away all burrs. Copper sheet of 24 s.w.g. will do well. A lip must first be turned up to prevent

Fig. 4. Mounting the lap in the lathe

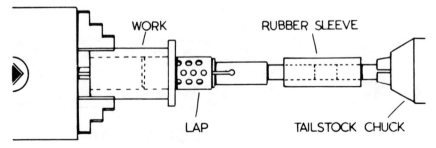

LAPPING

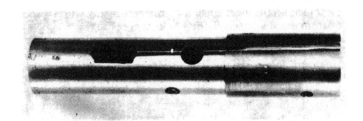

the lap turning in use after which the rolled up copper head is passed through a hole, or a series of holes if necessary, to size it.

Lapping Compounds

The production of lapping compounds is a specialist matter and not all makers of abrasives undertake it. The abrasive used is, for the most part, either aluminium oxide or silicon carbide held in suspension either in oil or water, and in some instances in grease.

Many grades are available ranging from 90 to 700 grit size. For the majority of work in the small workshop the finest grades will suffice; there is usually little metal to remove and time is not a significant factor. For this reason a compound of 500 grit can be used initially followed by one of 700 to polish the work surface.

The first compound can well be oil bound while the polishing agent is a water-suspended mix; and it goes without saying that both the lap carrier and the work need to be thoroughly cleaned when changing from one type of compound to another. In addition the lap itself needs to be replaced or the residual coarser compounds will .inhibit the polishing of the work surface.

Final polishing is commonly carried out with laps made of white wood, those for external work being held in the hand, followed by a cloth or felt pad.

Wood laps for internal work are usually made from dowelling, split axially for a short distance and expanded by card or 'wooden wedges'. No great pressure is needed, the work being run at a fast speed.

The subject of lapping is an extensive one; so it is only possible to sketch in an outline here. The choice of the correct compound for the work in hand is of the utmost importance, but here again it has been possible only to deal in generalities. Readers are therefore advised to consult abrasive manufacturers who are always ready to provide help and advice.

Protection of Machines when Lapping

Reference has already been made to the necessity of protecting the lathe and any other machines used in lapping operations.

When the lathe is employed all slides must be kept covered, preferably with rags held in place by some of the small permanent magnets available at most tool shops, a covering of newspaper over the rag helping to ensure that any excess of lapping compound does not penetrate through to the lathe bed itself.

The bore of the chuck must be well stuffed with newsprint or rag, the work, where possible, being held in the jaws away from the scroll so that rag may be introduced to protect it.

The tailstock, and the lever feed mechanism necessary to make the best use of the tailstock, scarcely need much protection as all the mechanism is well away from the area in which the compound is being employed. A rag thrown over the tailstock, however, will do much to minimise the inadvertent introduction of abrasive carried on the hands of the operator.

The drilling machine is a very suitable machine in which to carry out lapping operations. It has all the necessary mechanism for executing the reciprocating motion of the spindle carrying the lap; in addition the protection of the table presents no problems as this can be solved by a piece of newspaper placed over the table before any work is secured to it where necessary.

Moreover, when work has to be held in the hands, for the most part, the spindle can be locked against vertical movement, the work table swung away, and any excess lapping compound can then fall on to rag or newspaper placed on the bench.

An old pair of gloves should be kept to protect the hands during the process of lapping and kept with the equipment used, whilst all paper or rag used to protect the machines is burnt directly the operation has ended.

CHAPTER 14

Toolmakers Buttons

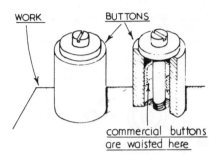

Fig. 1. Toolmakers buttons

WHILE it is possible to mark off a pair or a series of centres very accurately, if this condition is to be maintained in the actual machining operations, recourse must be had to 'toolmakers buttons' set up on the work to preserve this accuracy.

At one time sets of these buttons were available on the market. However, this lack is easily made good by fabricating the buttons for oneself, for they are simply hollow cylinders of steel whose end faces are square with their axes. They are secured to the work by screws, and may be moved about and then locked in place when correctly set.

A typical pair of toolmakers buttons are illustrated in *Fig. 1*.

It will be seen that ample clearance is given to allow positional adjustment to be made, while the washer used is of a size that will ensure that the button remains square with the work at all times.

They are used in conjunction with a dial indicator, the work being set on the faceplate, for example, and adjusted until the button is seen to be running true when the lathe is turned by hand. The buttons are usually made 0·300 in. or 0·500 in. in diameter and it is customary to make one of the set much longer than the others so that, when a pair of buttons is set up close on the work, as seen in *Fig. 2* the plunger of the indicator will not foul the second button.

Using and Setting the Buttons

As has already been said the buttons are attached to the work by means of screws, therefore, supposing for example that a pair of correctly meshing gears have to be mounted on a plate, the workpiece must first be marked off to indicate the centres for the two gears and then holes must be drilled and tapped at the intersection of the centre lines for the insertion of the screws securing the toolmaker's buttons. We will assume that the gears to be mounted are of 32 Diametral Pitch and have 16 and 32 teeth respectively.

Fig. 2. Setting buttons to run true

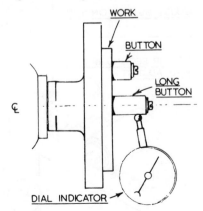

Their centres will therefore be $\dfrac{16}{2} + \dfrac{32}{2}$ of an inch apart or $\dfrac{8+16}{32} = \dfrac{24}{32} = \dfrac{3}{4}$-in. distant.

We will further assume that the first gear is to be positioned with its centre $\frac{1}{2}$ in. from both edges of the plate and that the second gear also has its centre located $\frac{1}{2}$ in. from the edge. The work must therefore, first be marked off as shown in the diagram *Fig. 3A*.

Fig. 3. Stages in setting buttons on work

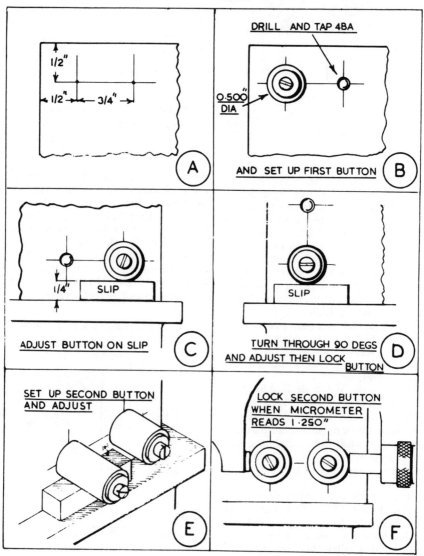

Two holes are then drilled, No. 32 drill, and tapped 4 BA to accept the screws that secure the buttons. The first of these is then mounted as shown in *Fig. 3B*. With the button in place the work is removed to the surface plate and stood on its edge. A parallel slip 0·250 in. square is then placed under the button which is tapped down until contact is made with the slip *Fig. 3C*. Next, the work is turned through 90 degrees and the button again tapped into contact with the slip. When this has been made the securing screw is finally tightened *Fig. 3D*.

The second and longer button is now placed on the work and the procedure used to set the first button is employed to ensure that both buttons lie on the same centre line *Fig. 3E*.

It now only remains to ensure that the buttons lie at the correct distance apart. Since the second button is now only held frictionally and can still be moved, a micrometer can be used to adjust the button in the manner depicted in *Fig. 3F*. When this has been done the button is locked in place, and the work removed to the lathe so that it can be set up on the faceplate.

The longest of the pair of buttons is the first to be set running true, using a dial indicator in the manner described earlier. When this has been done the button is removed and the work drilled and bored to whatever size is necessary. The second button is then set to run true and the procedure for drilling and boring repeated.

It must be emphasised that the example given is only one of the many ways of employing toolmakers buttons. In a book *Marking Off Practice*, once published by Percival Marshall & Co., the subject of these buttons was extensively covered. It is suggested that any readers who may need further information should enquire of local libraries if a copy of the book is still available.

CHAPTER 15　　　　　*Milling in the Lathe*

THE use of the lathe for carrying out milling operations has been well understood and practised for many years. Indeed, the earliest milling machines were, in the main, lathes adapted for the purpose.

The range of work possible is considerable, varying from the cutting of key-ways to the machining of gear wheels. Obviously, the magnitude of these operations is dependent upon the size of the lathe employed; and it is well, in this connection, to remember that only the utmost rigidity will result in satisfactory work.

Almost all specialised milling is dependent upon cutters produced commercially; these naturally require re-sharpening from time to time and this involves the use of equipment not normally available to the amateur. For this reason, where applicable, the process of flycutting is employed as this method avoids the use of specialised cutters and the attendant difficulty of re-sharpening.

Flycutting
Flycutting is the operation used in the lathe for producing machined surfaces on work held stationary while mounted on either the cross-slide or top-slide. Flat surfaces are obtained by rotating a single-point cutter mounted in an attachment that may be either held in the self-centring chuck or bolted to the faceplate. Curved work, and this is, of course, confined to concave surfaces, is produced by a cutter mounted in a bar normally rotating between centres. A typical piece of work here is the fitting of a chimney base to a locomotive boiler; the curvature of the mating face of the chimney skirt must agree exactly with the curved surface of the boiler itself so the chimney must be machined to the boiler curvature.

A boring bar with the cutter point set at the correct radius will suffice for this class of work; but, for flat surfacing, more rigid equipment is required. The simplest device for carrying the flycutter is that illustrated in *Fig. 1*. It consists of a Head 'A' carried on a Shank 'B' capable of being held in the chuck of the lathe. Two seatings are provided for the cutter; one parallel to the axis of the device, the other at some 45 degrees to it. The seatings are round in order to accommodate the circular section high-speed steel bits that are available as well as silver steel cutters that can be made for oneself.

The amount of material that may be removed at a single pass depends very naturally on the class of metal being machined; in the case of light alloy and brass and using a lathe of $3\frac{1}{2}$ in. centre height, 0·050 in. is quite a

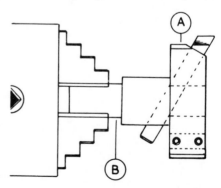

Fig. 1.　The flycutter

MILLING IN THE LATHE

Fig. 1. The flycutter

normal cut. Here the spindle speed employed is 400 r.p.m.

The flycutting of mild steel components, on the other hand requires a considerable reduction both in spindle speed and depth of cut. A maximum of 0·010 in. is usually as much as the process will stand with the spindle speed reduced to the slowest direct drive from the countershaft. The use of back-gear is not advisable; the intermittent tool loading inseparable from fly-cutting causes mechanical noise of an unpleasant and possibly damaging nature.

A modification of the device described, capable of carrying a pair of cutters, is that illustrated in *Fig. 2*.

The device is designed to be mounted in the 4-jaw independent chuck, and consists of a block of mild steel provided with seatings for square section ground tool bits made from high-speed steel. Ideally these seatings should be broached to ensure that they are both parallel and square. But

Fig. 1A. Flycutter components

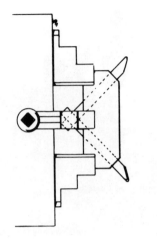

Fig. 2. Double flycutter

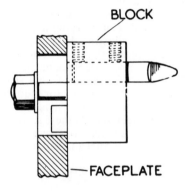

Fig. 3. Flycutter on faceplate

the amateur cannot be expected to have facilities for this, and so must fall back on the file to produce the seats after having first pilot drilled them.

Normally, both cutters are set out an equal distance from the block; but by adjusting the position of the block in the 4-jaw chuck, so as to make one cutter lead the other, and increasing the depth of cut of the lagging cutter the amount of stock removal per pass can be virtually doubled. The lathe faceplate is a very suitable mount for the flycutter for the reason that, when the tool is located upon it, the distance of its point from the headstock bearing, or overhang, is reduced to the minimum possible.

The support for the tool itself is very simple as it consists of a mild steel block, provided with a tenon to engage one of the slots in the faceplate, having one or more hexagon screws to secure the device in place. The arrangement is shown in the illustration *Fig. 3*.

If thought necessary the tool may be set at an angle; but, with the large area of work that may be covered by a fly-cutting tool mounted on the faceplate, an angular setting is hardly needed and is probably best omitted.

In addition to surface machining the flycutter may be used to cut keyways. The cutter used is similar to a parting tool employed in turning operations. It is mounted in a bar revolving between centres or caught in the 4-jaw chuck with the lathe tailstock brought up in support.

The work is fed against the direction of the cutters rotation as a precaution against tool-grabbing that must inevitably occur in a light machine, when the cutter is allowed to 'climb' the work. The technical term for feeding the work in the same direction as the rotation of the cutter is 'climb-milling'. The process has useful industrial applications, but needs machines of much rigidity having slides of sufficient weight to overcome the tendency the tool has of clawing the work towards itself.

The correct method of feeding the work is illustrated in *Fig. 4*.

When the cutter bar is mounted between the lathe centres considerable stiffening of the set up is afforded by using the fixed steady in support. The tool bits used for flycutting are, for the most part, similar to those used for turning; as such there is no problem involved in resharpening them by off-hand grinding. If necessary they are readily made from silver steel and are subsequently hardened and tempered. But for preference round or squared section high-speed steel tool bits should be used as the edge imparted to them by grinding and will stand up better to the rather arduous conditions of flycutting.

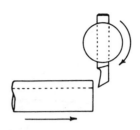

Fig. 4. Feeding the flycutter

Fig. 4A. Flycutting a long keyway

Fig. 5. Calculating packing

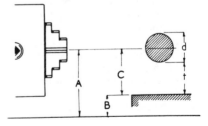

Mounting the Work

Whilst the ideal method of holding the work for end-milling is in a vertical slide, and reference has already been made to this, work may be clamped on the lathe top slide and packed to the correct height. This procedure has the merit that it enhances rigidity, though packing the work needs some care.

In this connection the correct thickness of packing required when end-milling a keyway in a shaft can be calculated with reference to the diagram *Fig. 5*.

First, the dimension 'C' must be established. This is obtained by subtracting the dimension 'B' that is the distance from the lathe bed to the surface of the top slide, from the dimension 'A' the lathe centre height. Both dimensions 'B' and 'C' must be measured accurately, preferably with a height gauge, otherwise the dimension 'C' will not be reliable.

Dimension 'A' is obtained by applying the blade of the height gauge to a mandrel held accurately in the 4-jaw chuck having been set to run true with a clock gauge. The required measurement is then obtained by subtracting half the diameter of the mandrel from the height gauge reading. The dimension 'B' is obtained by means of a direct reading from the height gauge applied to the surface of the top slide.

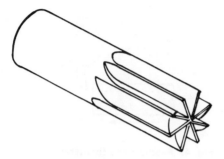

Fig. 6. The end mill

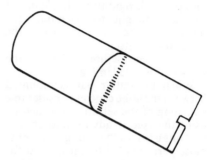

Fig. 7. The slot drill

When established, the dimension 'C' becomes a constant for the particular lathe being measured. It is then a simple matter to calculate 't' the amount of packing needed to mount the work truly on the lathe centre line.

Mounting Work on the Cross Slide

Work may also be mounted on the cross slide. Location of work in this way has the advantage of securing the maximum rigidity possible in the circumstances. Flycutting is best carried out when the work is bolted to the cross slide, the more so when the area of the work is considerable.

After marking off, setting the work for machining will, for the most part, be a matter of using the surface gauge to ensure that the job is correctly aligned. Again, recourse may have to be had to the dial test indicator to ensure that the work is square with the axis of the cutter.

Using End-mills

The End Mill illustrated in *Fig. 6* is commonly used for a variety of purposes such as the forming of hexagons and squares on small machined details. This type of cutter is also sometimes employed for cutting seatings for rectangular keys. Though the Slot Drill, a variant having two cutting lips only and illustrated in *Fig. 7*, is now more commonly used. Both the end mill and the slot drill illustrated need to be centrally mounted if they are to perform in a satisfactory manner.

For many operations it will be sufficient to hold these cutters in the self-centring chuck, but for work such as the cutting of keyways a little consideration will show that unless the slot drill turns quite truly the keyway will be machined oversize. The simplest way to ensure the necessary accuracy is to hold the cutter in the 4-jaw independent chuck. In this way it will be possible to use a dial test indicator to test the truth of the slot drill's rotation.

When cutting keyways the work should be fed to the cutter in one direction only, otherwise the keyway may be made oversize. In addition a start should be provided for the cutter by drilling a pilot hole to the full depth of the finished keyway. The pilot drill should be some 0·015 in. smaller than the slot drill in order to secure a good finish to the work.

Using Circular Saws

Circular saws are used in milling for a variety of purposes. In the lathe they can be used for slotting screws, for slitting the collars employed in shaft location or for cutting the slots in collet chucks.

MILLING IN THE LATHE

Essentially, saws are thin plain milling cutters. The number of teeth in them varies. For the most part fine toothed saws are only suitable for relatively shallow cuts, whereas saw with a comparatively few widely spaced teeth find their application in deep cutting. The reasons behind this are the better swarf clearance and lower power consumption of the widely spaced tooth.

Fig. 8 shows the tooth proportions of typical fine and coarse saws. These may be obtained in a variety of sizes, but for use in the amateur shop saws from 2½ to 3 in. in diameter with thicknesses 0·015 to 0·060 in. will be found sufficient for all average requirements.

In order to lessen drag in the work and to break up chips that might otherwise marr it, some saws are provided with side cutting faces. Naturally they have a greater thickness than those of the plain variety, for this reason their application in the lathe is somewhat limited. A typical example is shown in use in *Fig. 8a*. When used in the lathe, metal slitting saws are normally mounted on an arbor carried between centres. These arbors are similar to those fitted to a normal horizontal milling machine, and are ground on their own centres so that they will run truly when again so mounted.

The arbors are provided with distance collars enabling any cutters used to be placed in the most convenient position in relation to the work. The collars have to be made with the greatest care to make sure that their abutment faces, that is the faces that make contact with the sides of the cutter, are truly square with the axis of the arbor. As there are several of the distance pieces on the completely assembled arbor a little thought will show that an accumulation of errors here will result in a bent arbor directly the cutter is firmly secured by the lock nut.

The component parts of a typical arbor assembly are illustrated in *Fig. 9* Here, it will be seen, the collars are shown with their abutment faces

Fig. 8. Fine and coarse saw teeth.

Fig. 8A. Supporting a staggered tooth side and face cutter by means of the fixed steady

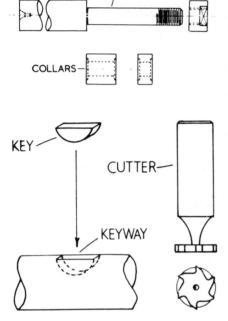

Fig. 9. The milling arbor

Fig. 10. Machining keyways

recessed. This allows them to be the more readily cleaned before assembly with the cutter to be used. Cleanliness here is essential, since the inclusion of metal particles between the mating faces will have an adverse effect on the alignment of the arbor itself.

It is sometimes convenient to catch the arbor in the 4-jaw independent chuck, setting it to run by means of a dial test indicator, and bring up the tailstock as support. In addition the fixed steady may be used to provide additional rigidity when needed.

Metal cutting saws are sometimes mounted on short arbors designed to be held in the chuck with or without tailstock support as occasion demands, but for the most part, their use is in connection with milling attachments to be considered later.

Vertical Slides

The range of work that may be carried out by milling in the lathe is greatly extended by the use of fitments known as vertical or milling slides. These are, in effect, angle plates capable of being bolted to the cross slide but having one face, the vertical face, able to be moved under the control of a feed screw, in this way work attached to the vertical face, either by being bolted directly to it or held in a small machine vice attached to the slide, may be set in the correct position relative to the cutter itself. A typical example is the cutting of the seatings for Woodruff keys depicted in *Fig. 10*.

Two forms of vertical slide are available; the first a plain slide with vertical movement only, the second a compound slide where both the slide itself and its base can be rotated on their own axes through 360 degrees. A vertical slide of the second category finds application in cutting bevel gears where the gear flank needs to be fed to

MILLING IN THE LATHE

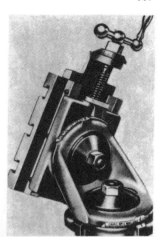

Fig. 11. Fixed vertical slide

Fig. 12. Compound vertical slide

Fig. 13. The Myford dividing head

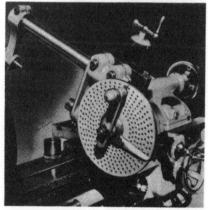

the cutter at an angle to the axis of the finished gear.

Typical examples of both categories of milling slide are illustrated in *Figs. 11 and 12* respectively. These slides are manufactured by the Myford Engineering Company. The simple slide is attached to the lathe cross slide by a pair of T-slot bolts whilst the compound slide has one centrally placed bolt set in the base of the slide mounting.

When gear cutting is to be undertaken some method of indexing the work is needed. The Myford Engineering Company supply a Dividing

Fig. 13A. Using the Myford dividing head

Fig. 14. Milling attachment
Fig. 15. Cutting a gear with the milling attachment

Head adapted for mounting on either of their milling slides. This dividing head is provided with mountings for chucks, and arbors may be set in the hollow mandrel fitted to the device which is illustrated in *Fig. 13*. The whole question of dividing in the lathe will be dealt with in detail later.

Milling with Lathe Attachments

An alternative to using the lathe headstock for driving milling cutters is the independently driven milling attachment as illustrated in *Fig. 14*. The device is intended for mounting on a milling slide and is fitted with back gear so that milling saws may be used and run to the best advantage. It is driven either from the lathe overhead or from a dwarf countershaft and low-voltage electric motor attached to the end of the cross slide.

The spindle nose of the attachment is the same as that used on the mandrel of the Myford ML 7 lathe; in addition the spindle itself is drilled axially and is bored No. 2 Morse Taper so that standard Myford Collets can be used as well as any peg-mounted drill chucks that are available.

An alternative milling attachment is illustrated in *Fig. 16*. Here a commercially made milling spindle and its driving motor are seen both mounted together on a vertical slide so that only a single short endless rubber driving belt is needed. This particular attachment runs at a high speed and is suitable for use with cutters up to a maximum diameter of $\frac{1}{4}$ in. The spindle of the attachment is bored No. 1 Morse Taper and is drilled for a draw rod enabling cutters with taper shanks to be mounted securely.

Driving Milling Attachments

Milling attachments of one form or another have been known for many years and the method of driving them has been the overhead lathe countershaft of more or less complicated layout. The problem, here, is the provision of a drive that will allow the milling attachment to be used any-

MILLING IN THE LATHE

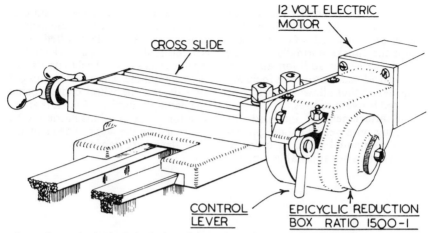

Fig. 18. Cross slide self-act

where along the bed of the lathe without running into trouble with driving belts coming off their pulleys. The advent of the electric motor, however, has done much to solve these problems by permitting the driving unit to be mounted either on the attachment itself or at the back of the cross slide as illustrated in *Fig. 18*, integrally with the milling spindle, as we have seen, or separately on the cross slide as depicted in the illustration *Fig. 13*. For further details of suitable belt drives reference should be made to Chapter 1.

Where the work is to be mounted on the saddle some means of holding and indexing it has to be provided. The most convenient way of satisfying this requirement and at the same time providing the greatest versatility, is to make use of a milling attachment similar to that illustrated in *Fig. 14*. This device is made by the Myford

Fig. 16. Milling attachment with self-contained motor

Fig. 17. Elements of a simple saddle mounted dividing device

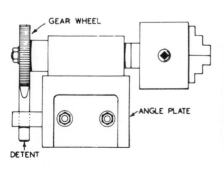

Engineering Company for use on their lathes, and is fitted with a hollow spindle provided with a nose that will accept the chucks normally supplied with the lathe. In addition an overarm with adjustable centre is included in the device thus allowing work to be mounted between centres.

The attachment is intended for mounting on a vertical slide, preferably a compound slide, in order to furnish a rigid but adjustable support. It is possible to make use of simpler devices, and in the main these usually consist of an angle plate bolted to the lathe cross slide having a work mounting allowing a gear wheel to be attached to the spindle. The wheel is engaged by a detent, the wheel chosen being one having a number of teeth that are a factor of the divisions needed to be machined on the work. The elements of the device are seen in the illustration *Fig. 17.*

CHAPTER 16   # Dividing in the Lathe

MANY of the milling operations carried out in the lathe, or indeed elsewhere, involve the use of some means of dividing the work. Examples are gear cutting, the forming of hexagons or squares on components and other work of a light machining nature. Dividing may be defined as the use of the machine to reduce a given piece of work into a number of equal parts.

In the lathe dividing may be employed either circumferentially when, for example, a ring of equally spaced holes needs to be marked out on a workpiece, or in a linear form as in making a scale or the machining of a rack.

If the lathe is to be used for milling purposes employing a cutter spindle secured in a fixture mounted on the saddle, then hexagons, squares and other regular shapes may be machined on work held by the lathe headstock; work held in this way will require the lathe itself to be used as a dividing engine.

Circumferential Dividing without Complex Attachments

Lathes are provided with a set of change wheels for screw cutting purposes. These wheels have an accuracy sufficient for all practical purposes so they can be used with advantage in a number of ways to carry out a dividing operation. All that is needed additionally is a detent or stop to engage the gear teeth at predetermined intervals and hold the work against rotation while the marking off or machining is carried out.

Much simple dividing can be performed using a single change wheel mounted on the stud of the lathe change wheel set up. This is equivalent to mounting the change wheel directly on the tail of the mandrel since both rotate at the same speed despite the possible interposing of a tumbler reverse gear. But care must be observed to ensure that all backlash is removed from the simple gear trains involved.

For the most part, when milling in the lathe, the number of divisions required are few. There are four divisions for squares, six divisions for hexagons and eight for octagons when needed. It is necessary, therefore, to choose wheels having numbers of teeth divisible by these numbers. The first of the divisions mentioned as well as the second can be satisfied by a 60-tooth wheel whilst the last conveniently makes use of a wheel having 40 teeth.

The detent previously referred to is bolted to the change wheel bracket and engages spaces between the wheel teeth as seen in the illustration *Fig. 1*. Details of a detent suitable for the Myford ML 7 lathe and the Drummond 3½ in. lathe are given at 'A' and 'B' respectively in *Fig. 2*. The detent may be seen in the place at the lower

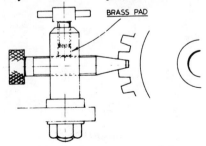

Fig. 1. Arrangement of detent

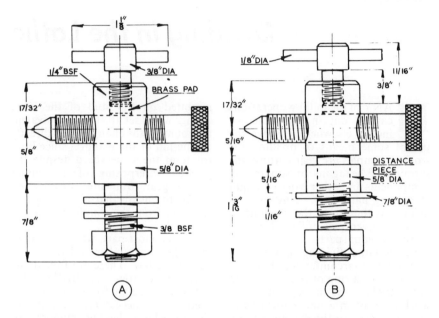

Fig. 2. Detent details

Fig. 3. Detent in place on Myford lathe

part of the illustration *Fig. 3*. The wheel is marked off with the necessary tooth spaces equally disposed. If, for example, a hexagon is to be formed then, with a 60-tooth wheel in use, every tenth tooth space will need to be noted and the detent engaged in each marked tooth space in turn. As has been said, the detent employed is usually mounted on the change wheel bracket, as this position allows it the more easily to engage the change wheel on the stud. The detent may take one of two forms, a simple screw-in device as illustrated in *Fig. 4* or the more advanced spring loaded arrangement seen in the illustration *Fig. 5*. The method of construction of the simple detent will be obvious. The pillar is provided with a screw lock and brass pad, enabling the detent to be locked firmly after registering with the change wheel.

The spring loaded detent is seen sectioned in *Fig. 6* and detailed in *Fig. 7*. It will, of course, be appreciated that the dimensions given are for a

DIVIDING IN THE LATHE

specific application; but their modification to suit individual requirements should present no difficulty.

Compound Dividing

We come now to a more advanced method of using change wheels for dividing. This employs a train of wheels and is used when a single wheel is not available. Let us suppose that it is necessary to divide a piece of work into 100 divisions; a common enough requirement, by the way, when making a feed screw index. If a wheel having 100 teeth was to hand no problem would arise since the detent could be used on every tooth space. For the most part, however, a 50-tooth wheel would have to be employed, this being the wheel normally supplied in the lathe change wheel set.

Fig. 5. Spring loaded detent

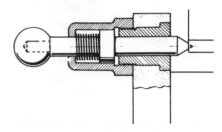

Fig. 6. Section of spring loaded detent

Fig. 4. Simple push-in detent

The set-up may be written down arithmetically as:

$$\frac{50 \times 40}{20} \text{ equals } \frac{2000}{20} \text{ equals 100 divisions.}$$

Similarly if there is a need for 125 divisions, sometimes used when making a leadscrew index, a 50-tooth wheel could be substituted for the 40-tooth thus giving:

$$\frac{50 \times 50}{20} \text{ equals } \frac{2500}{20}$$

equals 125 divisions.
Where no second 50-tooth wheel is supplied, however, it will be necessary to further extend the gear train; a

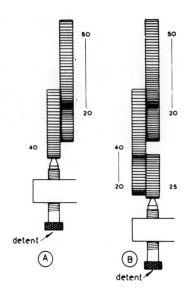

possible combination is set out arithmetically as:

$$\frac{50 \times 40 \times 25}{20 \times 20} \text{ equals } 125.$$

The wheels in the numerator are the drivers whilst those in the denominator the driven wheel. They are therefore connected in the following order:

Stud 50——20   25——Detent
        40——20

as seen in the illustration *Fig. 8* at 'B'.

Eliminating Backlash

It will, of course, be appreciated that in any train of gears there must be backlash resulting from the necessity of providing clearance between the teeth of the individual gears themselves. If no steps are taken to eliminate this backlash the resulting dividing of the work is liable to be inaccurate and of little account.

One simple way is that shown in the illustration *Fig. 9*. Here, a chuck key is inserted in the chuck holding the work

Fig. 8. Arrangement of wheels for compound dividing

Fig. 7. Details of spring loaded detent

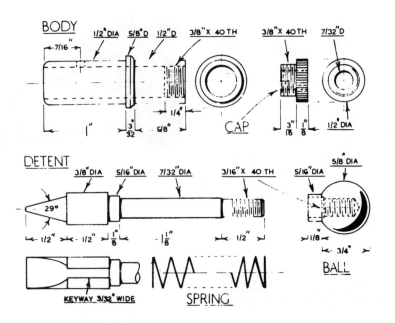

DIVIDING IN THE LATHE

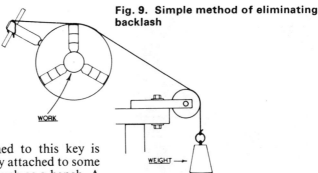

Fig. 9. Simple method of eliminating backlash

and a cord attached to this key is passed over a pulley attached to some convenient object such as a bench. A weight, of sufficient mass to ensure that the work will remain loaded against the gear train, is then tied to the end of the cord. An arrangement such as this, though simple, has in the past been found adequate when engraving the index dials of feed screws.

A Change Wheel Mounting for the Lathe Mandrel

When dividing from a single wheel, and to a lesser degree when a train of wheels is in use, backlash may be eliminated if the wheel can be mounted directly on the tail of the mandrel itself. This is most conveniently accomplished by making use of an expanding extension of the type shown in *Fig. 10* where a device designed and made for the Myford ML 7 lathe is illustrated. The extension fits into the end of the mandrel and is expanded by a tapered bolt so that it grips the mandrel. At the same time the change wheel mounted on the outer end of the extension is firmly held in place as a result of its own seating being also provided with means of expansion. Details of the device are given in *Fig. 11*.

A detent can now be made to engage a change wheel so mounted as seen in the illustration *Fig. 12* where it will also be apparent that a special fixture is needed to carry the detent.

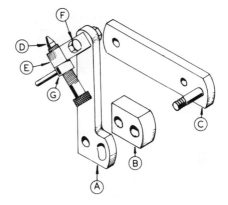

Fig. 13. Parts of the detent bracket

Fig. 10. Mandrel extension for the ML7

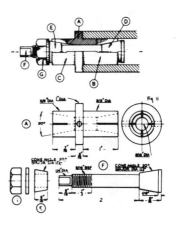

Fig. 11. Details of the mandrel extension

The parts of this fixture are depicted in the illustration *Fig. 13* and detailed in *Fig. 14*.

Dividing with Simple Attachments

A simple solution to dividing in the lathe is to make use of the bull wheel, that is the larger gear wheel attached to the mandrel itself. This has the advantage that work held in the chuck will remain fast and can be machined without fear of movement.

Unfortunately, many bull wheels do not readily lend themselves to this

Fig. 12. End of ML7 showing change wheel and detent

Fig. 15. Detent for the Drummond bull wheel

Fig. 14. Details of the parts

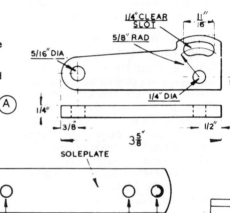

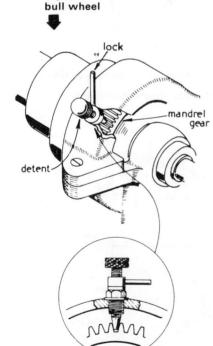

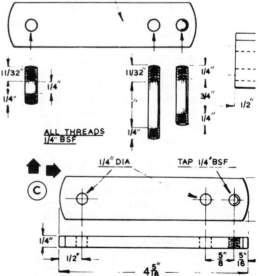

DIVIDING IN THE LATHE

purpose for the very good reason that the number of their teeth is unsuitable The 3½ in. Drummond lathe, for which the simple detent depicted in the illustration *Fig. 15* was made, has 66 teeth in the bull wheel so it is only possible to use it for hexagons or two flats diametrically opposite to each other

Simple indexing, for the most part, consists, as has been stated, in dividing work into two, four or six divisions. One way of providing for this is by drilling a ring of equally spaced holes in the chuck backplate. A detent may then be bolted at some convenient location to engage the holes as required. Such an arrangement is rigid and simple to use and is illustrated in *Fig. 16*. If the drilled holes are numbered from 1 to 12 indexing will be made the easier because it will not be necessary to actually mark the holes needed for any particular dividing operation. For example, when forming a square needing four divisions the detent will engage number 1, number 3, number 6 and number 9 holes successively and these figures can readily be memorised.

An alternative to this arrangement is the drilling of the face of the bull wheel itself as seen in *Fig. 17*. This

Fig. 14. (continued)

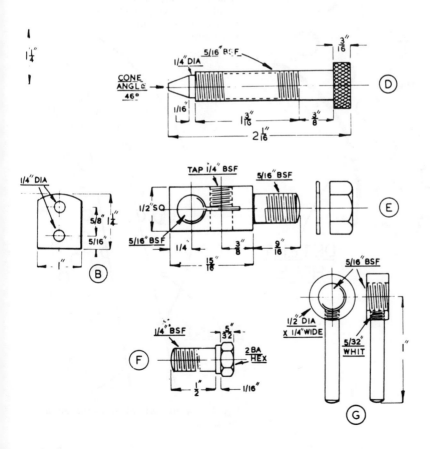

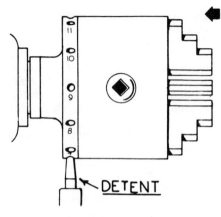

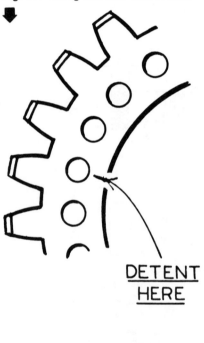

Fig. 17. Using the bull wheel face

Fig. 16. Using the chuck backplate

method offers the advantage that any equipment mounted on the mandrel nose will be held firmly during any machining process requiring indexing.

Many light lathes of a century and more ago, and in particular ornamental turning lathes, were provided with complete division plates attached either to the mandrel pulley when no back-gear was fitted or to the bull wheel itself when the lathe had a back-gear. These division plates had several rows of holes so the range of dividing possible was quite extensive. *Fig. 18* depicts their fitment. The modern lathe, however, with complete guarding of the moving parts on the headstock does not lend itself well to this treatment.

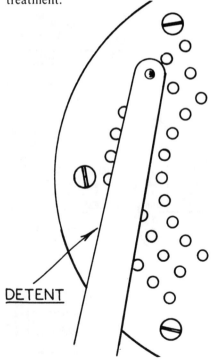

Fig. 18. A bull wheel mounted dividing plate

CHAPTER 17

Dividing

Dividing Attachment for the Headstock

We have already described the simple pieces of equipment that may be used for dividing directly from the headstock and it will have become apparent that, in order to carry out the more complex divisions sometimes needed, more advanced apparatus is needed. Such a device is illustrated in *Fig. 1*. This makes use of a change wheel mounted on the tail of the mandrel employing the expander extension previously described.

Before embarking on a description of the apparatus it will be well to understand the principle upon which a dividing head works. For the most part, dividing heads used industrially comprise basically a worm engaging a worm-wheel having 40 teeth (and sometimes 60). These elements are seen in *Fig. 1B*. The worm may be turned by hand enabling the operator to advance the wormwheel by increments as required by the number of the divisions to be made. One turn of the worm advances the wormwheel by one tooth so, by dividing the number of teeth in the wormwheel by the number of divisions required, it is possible to arrive at the number of turns needed, thus for example:

$$\frac{\text{Teeth in wheel}}{\text{Divisions needed}} = \frac{40}{20} = 2 \text{ turns}$$

Fig. 1. Lathe dividing head

Fig. 1B. Parts of the lathe dividing head

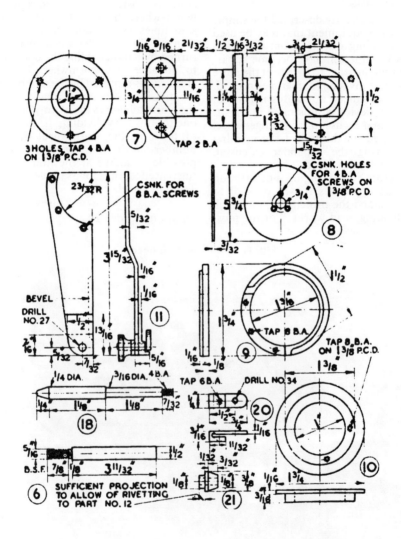

DIVIDING

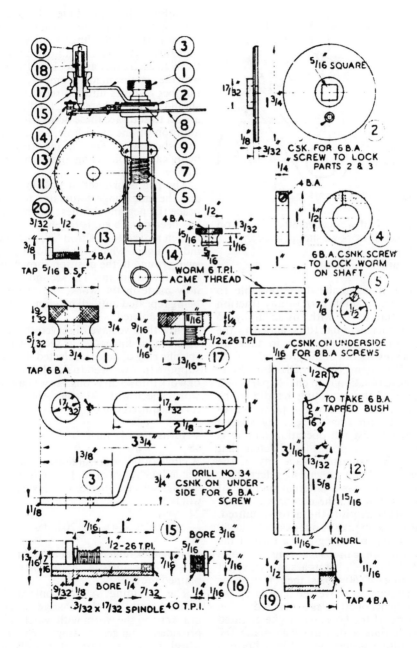

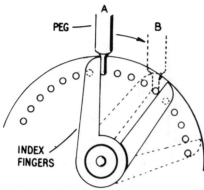

Fig. 2. Index fingers

Similarly if we needed 80 divisions on a piece of work we should need:

$\dfrac{40}{80}$ turns. i.e. $\dfrac{1}{2}$ a turn

Now it is manifestly impossible to estimate a single turn or any fraction of a turn by eye with any accuracy, so a further item has to be added to the dividing head in order to take all guess-work out of the process.

This item is the Division Plate whose function is to enable parts of a turn of the operating handle to be estimated. The division plate is provided with rings of equally spaced holes whilst the operating handle is fitted with a spring loaded detent that may be made to engage any of the rings of holes previously referred to. Each ring has a different number of holes in it and the numbers used are related directly to the factors required. If we again take as our example the 80 divisions requiring half a turn of the worm we can use a ring of holes divisible by 2. Let us assume we select a ring of 60 holes then it will be necessary to move the handle by increments of 30 holes for each division made. If the holes had to be counted each time a division was made the resulting process would be laborious and time-wasting. So a further piece of equipment is added, this is the assembly called the Index Fingers. This assembly rotates around the wormshaft and the fingers, of which there are a pair, can be adjusted relatively to each other and locked. In this way they can be made to embrace any number of holes interval required. It should be noted here that the fingers are set to embrace these holes with the addition of one extra hole, as seen in the illustration *Fig. 2*, indexing is being carried out using an interval of 4 holes.

The peg is first set in position 'A' and the index fingers adjusted to embrace the peg and 4 holes. When this has been done a cut is taken. The peg is then moved to position 'B', and the index fingers, which are friction tight, swing round to the position shown by their dotted outline. The second cut is now made in the work, the process being repeated until all machining has been completed.

Now that we have observed the use of the index fingers let us see how the specific ring of holes is selected. Let us suppose that we need to divide a piece of work into 48 divisions. As we have a wormwheel with 40 teeth then the turns required will be:

$\dfrac{40}{48}$ turns or $\dfrac{5}{6}$ of a turn

A circle of holes divisible by 6 will be needed such as 24 or 36. If a circle of 24 holes is selected then 5/6 of this circle will be:

$\dfrac{24}{6} \times 5$ holes $= 4 \times 5 = 20$ holes

So the index fingers are set to embrace $20 + 1 = 21$ holes in the manner explained previously.

The Dividing Attachment itself bolts directly to the change wheel bracket and the worm with which it is provided engages a single change wheel since no train of wheels is needed. No allowance is made for the helix angle of the worm which has a pitch of $\frac{1}{16}$ in., but if thought desirable this angle could be compensated for by introducing a wedge packing piece between the change wheel bracket and that supporting the attachment. The device has been designed so that when using change wheels of 20 diametral pitch one turn of the attachment handle advances such a change wheel one tooth. If reference is made to the illustration *Fig. 1B* it will be seen that the apparatus consists initially of a bearing assembly carrying the worm shaft. This assembly is attached to the change wheel bracket. An extension of the worm shaft carries the operating handle and spring detent whilst the division plate itself is attached to the top of the bearing housing. The fingers for the division plate are located on its upper surface and rotate around a hub carried on the wormshaft. The fingers may be set and clamped together, and also secured to the division plate after each dividing movement has taken place.

Dividing Attachment for the Lathe Saddle

A dividing attachment may also be used on the lathe saddle. This enables the lathe to be used for gear cutting with the work mounted on the attachment and the cutter carried by the lathe headstock. It also allows the lathe to be employed, for positional drilling such as is required by a ring of holes on a given pitch circle. An example of this equipment is seen in the illustration *Fig. 3* where the Myford Dividing Head is depicted. This device is designed to be attached to a vertical slide mounted on the lathe cross slide. It is provided with an overarm and centre to support work either mounted on centres or held in a chuck attached to the spindle. This spindle is bored the same size as that of the Myford lathe and carries a No. 2 Morse Taper so that a normal lathe centre may be used; the spindle nose is a replica of that found on the lathe

Fig. 3. The Myford dividing head

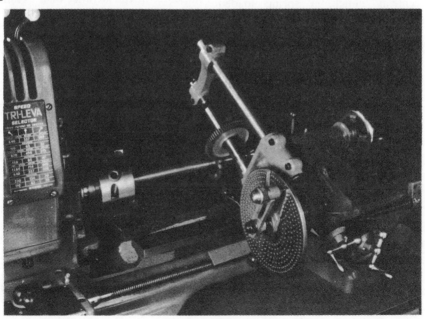

itself enabling the standard chucks, driver plate and face plate to be mounted when required. In addition the Myford range of collet chucks may be used with the attachment.

Simple Dividing Attachment for the Lathe Saddle

A simple device for indexing and capable of being used with a vertical milling slide is that shown in the illustration *Fig. 4*. It consists of a base plate with an attachment for a self-centring chuck, a spring loaded plunger detent and a means of locking the chuck against rotation after each indexing movement.

As the device is only intended for simple dividing the rim of the chuck back-plate is drilled with 12 equally spaced holes. This enables indexing for 2, 3, 4 and 6 divisions to be made rapidly, and will suffice for those divisions most commonly needed.

Fig. 4. Simple dividing attachment for the lathe saddle

The attachment was made, in the first instance, to permit index drilling of small components to be carried out. But, subsequently, it has been found useful in the rapid end-milling of small batches of components requiring spanner flats and in other work of a like nature.

The details of the device are seen in the illustration *Fig. 5*.

Linear Dividing

An extension of the use of the lathe as a dividing engine is its employment for the purpose of cutting racks, engraving scales and work of a like nature.

These operations make use of the leadscrew as the necessary medium, the process varying from the simple to the more complex according to the nature of the work involved.

If the leadscrew is provided with a micrometer index wheel it will clearly be possible to use this fitment to measure off the advance of the leadscrew required for any particular piece of work. But such a practice is likely to be time-consuming and may introduce inaccuracies if, for any reason, the reading of the index has not been correct. It is best, therefore, to employ the methods of the milling machine operator and set up a simple wheel train to drive the leadscrew and make certain that the dividing is accurately carried out.

An example of the set-up required is illustrated in *Fig. 6* and diagrammatically in *Fig. 7*.

It consists of a change wheel bracket carrying a detent engaging a control wheel having one tooth space only. This wheel is coupled to a change wheel which in turn is meshed with a further change wheel, the ratio between the two wheels being such that one complete turn of the control

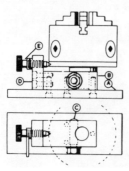

Fig. 5. Details of above

Fig. 6. Set-up for linear dividing

Fig. 7. Diagram of above

wheel advances the leadscrew by the amount desired.

This is analogous to the process of screw-cutting so the calculations needed to set up a train of wheels for dividing are similar.

A simple example will suffice to show what is involved. We will assume that it is desired to engrave a scale into divisions of $\frac{1}{16}$ in., a procedure suitable for engraving the barrel of a lathe tailstock to fit it as a depth gauge. We will also assume that, as in the case of the Myford lathe, the leadscrew has eight threads to the inch, i.e., $\frac{1}{8}$ in. pitch. If we are to advance the engraving tool by increments of $\frac{1}{16}$ in. then the ratio between the control wheel and that attached to the leadscrew would be:

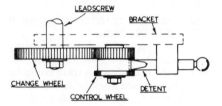

$\frac{8}{16}$ — or as 2:1

This is a ratio that is, of course, easily satisfied by a pair of wheels found normally amongst those supplied as standard with a lathe, a 60 wheel mounted on the leadscrew and a 30 coupled to the control wheel are

Fig. 8. Diagram for linear dividing

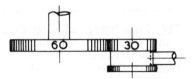

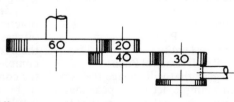

Fig. 9. Diagram for linear dividing

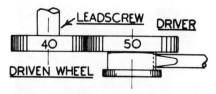

Fig. 10. Set-up for rack cutting

obvious examples as depicted in the diagram *Fig. 8.*

If, on the other hand, it had been necessary to have graduated the tailstock by increments of $\frac{1}{32}$ in. then the ratio of the wheels would have been:

$$\frac{8}{32} = \frac{1}{4} \text{ or as } 4:1$$

a ratio satisfied by the combination of a 20 and an 80 wheel.

In the event of an 80 wheel being unavailable, a compound train will have to be used. Such a train could be the wheels used for $\frac{1}{16}$ in. divisions plus a further pair of wheels to give a further 2:1 reduction. Thus as seen in *Fig. 9.*

60—→20
—
40—→30←—DETENT

Cutting Racks

An operation that may be performed on the centre lathe is the cutting of racks. A rack may be described as a gear wheel whose teeth have been extended into a straight line. It follows then that the distance between the centres of adjacent teeth of the rack is the same as that of a gear wheel having the same pitch. This distance, known as the circular pitch, is conveniently expressed by:—

$$\frac{\pi}{P},$$

where $\pi = 3 \cdot 142$
and $P = $ Diametral Pitch
of the gear that will engage the rack. As an example, let us suppose that a rack has to be cut that will mesh with a gear wheel having teeth 20 diametral pitch. Then the circular pitch will be:

$$\frac{\pi}{20} = \frac{3 \cdot 142}{20} = 0 \cdot 157 \text{ in.}$$

The wheels will be geared together in the order shown by *Fig. 10*.

This arrangement provides sufficient accuracy for most purposes as will be apparent from the table below, where wheel combinations for many gear pitches are given:

D.P.	Circular pitch	Driver		Driven	Error in.
12	0·262	40	2	38	0·001
14	0·224	45	2	50	0·001
16	0·196	40	2	50	0·004
18	0·175	35	1	25	Nil
20	0·156	50	1	40	0·001
22	0·143	40	1	35	Nil
24	0·131	40	1	38	Nil
26	0·121	75	1	80	0·004
28	0·112	45	1	50	Nil
30	0·105	80	1	95	Nil
32	0·098	40	1	50	0·002
36	0·087	35	1	50	Nil
40	0·079	25	1	40	0·001

The single figures adjacent to the Driving Wheels are the number of turns these wheels have to make.

Whereas, as has been said, the foregoing system provides a ready means of indexing a rack to limits that are practically acceptable, certain of the pitches have errors that would unfit them for use in circumstances needing complete accuracy. As an example, the combination of a rack and pinion to be used for measuring purposes would need a high degree of accuracy.

DIVIDING

In order to secure the greatest possible accuracy it is necessary to bring into the wheel combination the factor $\pi = 3\cdot 142$. If we take this figure as represented by the fraction $\dfrac{22}{7}$ then wheels that will satisfy it are found in the following way:

$$\frac{22}{7} \times \frac{11\times 2\times 5}{7\times 5} = \frac{55}{35}$$

The 35 wheel is mounted on the leadscrew whilst the 55 wheel is placed on the stud and is geared to the 35 wheel itself as shown in *Fig. 11*.

If we assume that the leadscrew has a thread of $\tfrac{1}{8}$ in. pitch then the advance of the saddle and any work secured to it for each movement of one tooth of the 35 wheel will be:

$$\frac{125}{35} = 0\cdot 0035\,\text{in.}$$

One turn of the 55 tooth wheel geared to it will then be $55\times 0\cdot 0035$ in. $= 0\cdot 196$ in. which is the circular pitch of a gear of 16 D.P. It is not convenient, however, to use only the 55 gear for indexing purposes so an additional gear has to be mounted on the stud alongside the 55 gear to enable indexing to be carried out. The arrangement is seen in *Fig. 12*.

If for example a 60 toothed wheel is selected for indexing purposes then the advance per tooth is given by dividing the number of teeth in the index wheel into $0\cdot 196$ so:

$$\frac{0\cdot 196}{60} = 0\cdot 0032\,\text{in.}$$

Supposing, therefore, that it is a rack having teeth 20 D.P. that is to be cut, then the number of teeth in the indexing wheel that need to be used is given by:

$$\frac{0\cdot 156\ (\text{circular pitch 20 D.P.})}{0\cdot 0032} = 48\ \text{teeth.}$$

A.W.   K

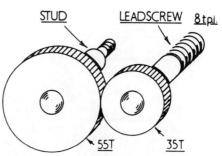

Fig. 11. Train of gears to satisfy the function π

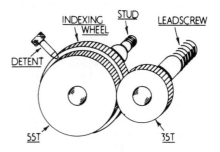

Fig. 12. Complete train of gears for accurate indexing

The error in this instance is only $0\cdot 00004$ in. in a total of 48 teeth, for all practical purposes a negligible figure, and it is this method that has been used in calculating the Table II giving figures for indexing a range of tooth pitches.

When indexing it is advisable to set a pair of dividers to embrace the number of tooth spaces necessary and to mark these in turn as they are used

Fig. 13. The rotary table

to ensure that the dividing operation remains in step.

Diametral pitch	Linear pitch	Indexing wheel	No. of teeth indexed
12	0·262	60	80
14	0·224	35	40
16	0·196	60	60
18	0·175	45	40
20	0·157	60	48
22	0·143	55	40
24	0·131	60	40
26	0·121	65	40
28	0·112	35	20
30	0·105	60	32
32	0·098	60	30
36	0·087	45	20
40	0·079	60	24

When the wheel train has been set up it is as well to check it. This is readily carried out by mounting a dial indicator on the lathe bed and using it to measure the movement of the saddle under the control of the index wheel. A series of consecutive measurements should be taken setting the indicator to zero between each movement of the wheel.

The Rotary Table

The last device for dividing we have to consider is the Rotary Table. Though not strictly of prime use on the lathe, being intended as an accessory to the milling machine, under certain conditions it nevertheless, may find employment there and so merits some attention.

Its basic mechanism is simple, consisting of a casting carrying the work table and a set of bearings to support the worm which engages a worm wheel forming part of the work table. The worm shaft is provided with a hand wheel for operational purposes and both it and the table may be locked against rotation under load.

The rim of the table is calibrated and is engraved with a scale of degrees. The main casting carries a zero plate, and in some instances a vernier to enable the finer angular measurements to be made. Additionally the operating hand-wheel is sometimes engraved with a scale, a development that somewhat simplifies the making of accurate measurement.

Fig. 14. Set-up for engraving on the rotary table

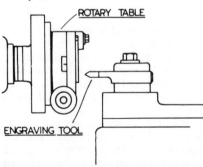

DIVIDING

The surface of the table is crossed with T-slots so that the work may be secured to it by means of bolts engaging the T-slots.

The centre of the table is bored accurately so that in addition to 'a dead centre' plugs or bungs may be used to centralise work or even, in some instances, to mount it.

In dividing work it will be apparent that all measurements are made in degrees or parts of them, this enables dividing to be carried out quickly and with certainty. Thus if a hexagon is to be machined the work is moved 60 degrees each time a face has been completed; if a square is required the movement is 90 degrees and so on according to the number of faces needed, these movements being determined by dividing the figure of 360 degrees by the number of faces desired.

Engraving a Protractor

An example of the class of work that can be carried out with a Rotary Table when mounted in the lathe is the protractor. In order to engrave it the rotary table is secured to the lathe faceplate. The work is secured to the face of the table with the engraving tool mounted on the lathe top slide as seen in the illustration *Fig. 14*.

The lathe mandrel is locked against rotation and all dividing increments carried out by the rotary table.

CHAPTER 18 # Drills and Drilling

IN most books written for beginners some consideration is usually given to the question of drills and drilling, and enough is normally said to enable the beginner to furnish himself with sufficient equipment in order to carry out work suited to his skill and the extent of his knowledge. As this knowledge increases he will naturally wish to extend his facilities, so it is the purpose of the present chapter to give some guidance on the subject.

At the moment of writing the desirability of 'going metric' is uppermost in everyone's mind. The value or otherwise of doing so commercially is a matter for debate, but whichever way the argument is settled it seems hardly sensible for the amateur who may have painstakingly collected together a set of drills, and for that matter screwing tackle based on English practice, to throw the whole lot away and start all over again. It should be emphasised, however, that of late years the extension of the available range of metric drills, especially in the smaller categories, has made them a valuable substitute for some of the drill sizes. Letter and number drills, for example, being well covered by the metric range.

The range advances by close increments, in point of fact a few thousandths-of-an-inch at a time, so the purchaser has available to him drill sizes that will be of the greatest help when either reaming or tapping threads.

The best advice we can give for the moment, therefore, is to retain the sets already in the workshop and to supplement these by individual metric drills as occasion demands.

At auction, or sometimes on the disposal market, it is often possible to acquire drills with tapered shanks made to the Morse standard. Apart from their obvious advantages when used in connection with drilling machines equipped to accept them, taper shank drills can usefully be employed on the lathe. A selected few of the larger sizes will be sufficient, but care should be taken to see that the drills chosen are not too large for the machines with which it is intended to use them.

In the case of second-hand equipment offered for sale the state of the tapers themselves should be well inspected before any drills are acquired. Many of the shanks get damaged in service so it is well to choose carefully before selecting a particular drill. Slight damage may be put right by polishing with an oil stone or a swiss file, drill shanks being usually left soft; but a severe case of bruising will need the surface to be machined away as depicted in the illustration *Fig. 1* at 'B'.

If it is intended to use taper shank drills in a drilling machine adapted for the purpose, the tangs of the drill themselves should be examined for damage. As has been shown elsewhere the tang fits into a recess formed in the drill spindle itself; it prevents the drill turning under load and has an essential part to play in the method used to remove the drill when required as illustrated in *Fig. 2*.

As will be seen a wedge, passed through a slot in the drill spindle, makes contact with both the tang and the spindle itself. A smart hammer blow upon the end of the wedge drives it in and ejects the drill.

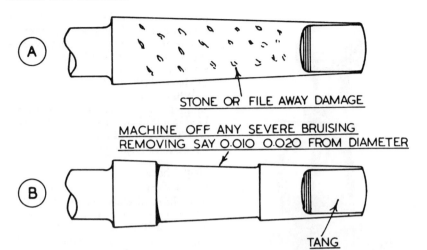

Fig. 1. Correcting damage to a taper shank

Drill Stands

It is always advisable to store sets of drills in stands. In this way not only can sizes be distinguished at a glance but the drills themselves will be better maintained than when all jumbled together in a box. While commercially produced drill stands are readily available, many workers prefer to make what is necessary for themselves, the more so when some special grouping of the drills is desired.

For example, as an examination of the handbook *Screw Threads and Twist Drills* will show, the percentage engagement of screw threads varies directly with the size of tapping drill used. Some workers, therefore, prefer, in connection with the screw threads habitually used by them, to group together the various tapping size drills they need.

The purchase of a drill stand for such a purpose may very well be a difficult, if not an impossible undertaking. But the want can easily be filled by a hardwood block, drilled at convenient spacings by those drills themselves that are to occupy the stand, the block itself being shellaced or french polished subsequently to prevent staining by oil.

Whilst it is possible to mark the drill sizes opposite the holes, using letter punches for the purpose, unless the drills are well spaced it is difficult to read the sizes. So, in the main, it is better to omit this refinement and rely on experience and visual skill in the first instance to select the drill sizes. In any event, and especially where the smaller drills are concerned, the diameter of those selected is best checked by means of a micrometer calliper.

It may not be out of place, at this point, to sound a note of warning.

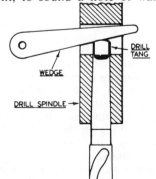

Fig. 2. The removal of taper shank drills

Fig. 3. Mounting work in a V-block

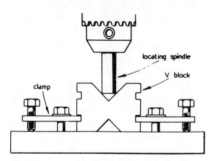

Fig. 3A. Locating the V-block

Possibly as a result of mechanical packaging, it does not necessarily follow that the size of drill inside a packet corresponds with the drill size marked on the outside. We recently had occasion to need some drills 4·3 mm diameter. These were duly obtained and put to use after checking, by the marking on the outside of the packet only, that the drills were as required. But it was not till the holes in a whole batch of work had been found oversize that it was discovered that drills 4·4 mm diameter had been placed in the packet by mistake and this by a firm of international repute.

The moral is obvious—check everything and do not take for granted the sizes marked on packets or stamped on the drills themselves.

When buying new drills in the smaller sizes it is also worthwhile having a look at the drill points themselves. It is not unknown for these to have been sharpened incorrectly.

Drill Chucks

Chucks for use with parallel shank drills have already been described in Chapter IV where equipment needed with the drilling machine is dealt with. Readers are therefore referred to the notes contained in that chapter.

Methods of Holding Work

There are several ways of holding work on the drilling machine table. The first, and unfortunately most obvious method, is to hold it by hand. But this, for the most part, is a dangerous practice and not to be recommended. If the part is large and heavy it may well stand up against the drill forces without being secured to the machine table. But the small components made by the amateur need to be firmly held in order both to protect the operator as well as to ensure accurate drilling.

A method commonly used to hold round work is illustrated in *Fig. 3*, where a short length of shaft is seen mounted in a V-block so that it can be located under the drill point and held firmly. The block itself is provided with grooves; the clamp securing the work and the dogs fastening the block to the drilling machine table engage these grooves holding the assembly firmly.

When work is mounted in a V-block for cross-drilling it is essential that it be set accurately or the drill will not pass directly through the centre of the work. A simple way to ensure that it

DRILLS AND DRILLING

does so is depicted in the illustration *Fig. 3A*. This shows a piece of round material gripped in the drill chuck being brought into contact with the V-block and held there while the clamps securing the block to the machine table are tightened.

It is important to see that, when the locating spindle is removed from the chuck, enough room has been left for the insertion of the drill to be used; for movement of the table after the V-block has been set correctly will undo all the accuracy of positioning.

Castings and other large components can be bolted or clamped to the table of the drilling machine while the smaller details are gripped in a machine vice. There is a number of suitable vices on the market; of particular service to the worker in the small workshop is the vice made by the Myford Engineering Company. This has a maximum jaw opening of $1\frac{1}{2}$ in. with a width of $1\frac{5}{8}$ in. and a depth of $\frac{3}{4}$ in. The vice is reasonably priced and is accurate, a requirement essential to the serious user. It is also of simple

Fig. 4. The Myford machine vice

Fig. 5. The Offen vice

design but it has all the essential elements of a good machine vice. An example is illustrated in *Fig. 4*.

Another small vice that may be used on the drilling machine is the Offen Versatile Vice illustrated in *Fig. 5*. This piece of equipment is provided with a base machined on its underside enabling it to be bolted firmly to the table of the drilling machine. The base carries a ball mounting to which the vice itself is secured; this allows the vice to be oriented and locked at any angle needed in relation to the base.

The vice, which is well made, has a capacity of 1 in., the jaws being $1\frac{5}{8}$ in. wide and the throat $1\frac{13}{16}$ in. deep. It will be observed that the screw is protected, a worthwhile improvement in a piece of equipment that may be used on the bench as well as on the machine.

The Drilling Machine Setting Ring
Fig. 6

When setting work on drilling machines not provided with mechanical means of raising or lowering the machine table, difficulty often exists

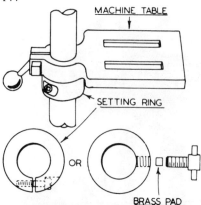

Fig. 6. The drilling machine setting ring

in positioning work under the drill. To do so it is sometimes necessary to swing the table around the column of the drilling machine. In doing so the table may slip downwards, possibly upsetting the depth gauging of the drill spindle. This may be avoided by fitting a setting ring below the casting supporting the machine table. With this ring clamped to the column the table may be swung around without endangering its vertical position. When providing this additional fitment it should be noted that a clamping ring is preferable to one that relies on a grub screw to secure it. If a grub screw is used a brass pad should be interposed between the screw and the machine column in order to avoid damaging the latter.

Depth Stops and Methods of Measuring Depth

When drilling 'blind' holes, that is to say holes that do not pass clear through the work, it is important to be able to ensure that the point of the drill only penetrates to the required depth. Whilst it may be possible, when drilling a single hole, to provide for this requirement by direct mensuration, but from the viewpoints of both time and accuracy a series of similar holes need a depth stop to make certain holes are all the same depth.

On some types of commercially made drilling machines the stop fixture consists of a clamp surrounding the quill of the drilling machine and carrying a threaded upright upon which two stop units run, the one acting as a lock for the other. The casting of the drilling machine is furnished with a drilled lug through which the threaded upright passes and it is against this lug that the stop nuts abut. All the parts that comprise the stop fixture are illustrated in *Fig. 7*.

The upright is marked off in a scale of inches but the actual operation of setting the depth of hole required is somewhat elementary and slow. So much so that we long ago decided to modify the arrangement in order to

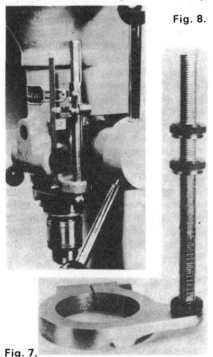

Fig. 7.

Fig. 8. Modified drilling machine stop

Fig. 7. Drilling machine stop

speed up the depth-setting process and to incorporate a steel rule that could be read more easily than the somewhat elementary scale engraved on the original upright. The modification is illustrated in *Fig. 8*. A cylindrical upright having an axial keyway now replaces the original one, and on this slides an adjustable stop having a bronze feather key engaging the keyway and preventing the stop from coming out of alignment. A thin steel finger is attached to the stop and this has an engraved index line that can be superimposed on the rule itself. The latter has a holder, screwed to the body of the drilling machine; this is provided with a clamp to secure the rule after its position has been adjusted in relation to the stop. In order to set the stop the drill point is first brought into contact with the work and the drill spindle locked. The stop is then lowered on to the lug on the head casting and a note taken of the reading shown on the rule which should be first set to some convenient figure. The stop is then moved away from the lug a distance equal to the depth of drilling required and is then locked to the upright, the amount of movement being read off on the rule. The drill spindle is then released and drilling commences.

Where it is possible to fit it the simplest way to provide a depth stop is to attach a clamp ring to the top of the drill spindle itself. The ring makes contact with the face of the driving pulley so can be set, after the drill point has been brought into contact with the work, by inserting a drill shank or a simple setting gauge between the two parts. The arrangement is depicted in the illustration *Fig. 9*.

The gauge is readily made from a length of flat material, the steps being arranged as shown. It is suggested that a maximum depth of $\frac{3}{4}$ in. is allowed for, increments of $\frac{1}{16}$ in. being set out starting from $\frac{1}{8}$ in. If intermediate

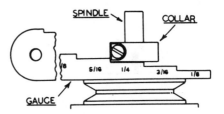

Fig. 9. Depth stop gauge

Fig. 10. Potts drilling grinding jig

depths such as $\frac{3}{32}$ in., $\frac{5}{32}$ in. and so on are needed then a second gauge can profitably be made.

Designers of drilling machines, in which the spindle runs in a quill having a rack cut upon it, sometimes make use of the rack pinion shaft to carry an engraved sleeve for measuring depth. The sleeve is friction-loaded so that, after the drill point has been brought into contact with the work, the sleeve is set at zero allowing the depth of the drilling to be read off directly.

Drill Grinding

Whilst it is possible to resharpen a drill by means of an off-hand grinding operation, by this method the possi-

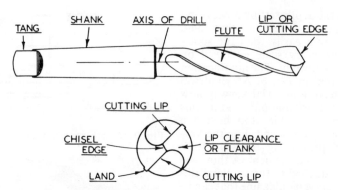

Fig. 11. The elements of the twist drill

bilities of ensuring that the tips are of equal length and at the correct angle to the long axis of the drill itself are almost nil. It is of importance that these requirements are met, if not the drilled hole will neither be straight, nor true to size.

A drill grinding jig will make certain that the drill point is accurately resharpened, enabling the user as it does to carry out this work expeditiously. One of the best known commercial fixtures obtainable in this country is the Potts Drill Grinding Jig seen in the illustration *Fig. 10*. The device is designed so that it can be bolted either to the bench alongside the grinding machine or directly to it, and is designed to enable drills to be sharpened on the side of the wheel.

The amateur, however, should not be actually discouraged from the practice of off-hand grinding. If he has no fixtures he must of necessity resort to it, experimenting initially with one of the larger drills, say one of ½ in. diameter. This will allow him the more readily to check the angle of the drill point as well as the length of the lips.

The Drill Point

Whatever method is employed to sharpen it, it is important to understand the elements both of the drill and the drill point and the nomenclature used to describe them.

The terms applied to these elements are set out in the illustration *Fig. 11*.

For general purposes the included angle of the drill point is usually made 118 degrees as seen in *Fig. 12*, whilst the lip clearance or back-off from the cutting edge normally has an angle of from 9 to 12 degrees as depicted in *Fig. 13* at A and B. If the clearance is increased materially beyond this

Fig. 12. Drill point angle

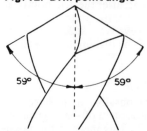

Fig. 13. Drill lip clearance

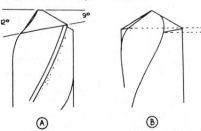

DRILLS AND DRILLING

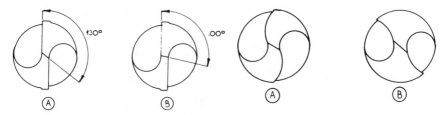

Fig. 14. Chisel point angle

Fig. 15. Faults in drill grinding

figure the cutting edge will be weakened and may chip during the drilling operation.

In a correctly ground drill the angle the chisel point makes with the lip or cutting edge is 130 degrees as seen in *Fig. 14* at A. This angle depends however, on the amount of back-off given behind the lip. If no clearance is given and the drill point ground like a lathe centre the chisel point angle would be reduced to 110 degrees according to the illustration *Fig. 14* at B.

Two common faults in drill grinding are illustrated in *Fig. 15* at A and B respectively. In 'A' the back-off given is too great and, in consequence, the cutting edges have become thinned and weak, as well as crescent shaped.

In 'B' the included angle of the drill point has been made greater than the standard 118 degrees resulting in the cutting edges becoming hooked.

From all of this information it will be clear that, when grinding by off-hand methods, the chances of producing a theoretically correct drill point are extremely thin. The drill manufacturers all stress this, but concede that some experienced operators do occasionally produce an accurately sharpened point free-hand. Those who wish to experiment with this method of grinding would do well, in the first instance, to set up a surface plate on its side as depicted in *Fig. 16*. The plate can then be smeared with marking blue and a correctly sharpened drill point rolled against it. In this way, by observing the amount of blue transferred to the drill point, the operator can check the effectiveness of his movements, and whether these will be successful when he makes use of the grinding wheel.

Drills are usually ground on the side of the wheel as depicted in the illustration *Fig. 17*. This leaves the

Fig. 16. Testing free hand grinding methods

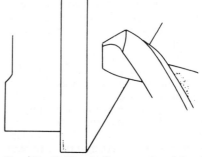

Fig. 17. Drill grinding on the side of the wheel

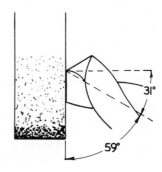

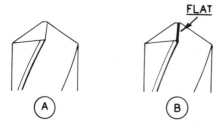

Fig. 18. Modified drill point for brass

worker with an additional problem, that of maintaining the correct angle of inclination, namely 59 degrees, when using freehand methods. However, if a jig is employed, no such difficulty arises for the angle is already set and is incorporated in the basic geometry of the design.

When the point of a correctly sharpened drill is examined it will be seen that the cutting edge has considerable top rake in the manner imparted to a lathe tool intended for cutting steel.

This condition is depicted at 'A' in *Fig. 18*. When a drill sharpened in this way is used for cutting brass it will be found to 'grab' or 'dig' in. This tendency can be overcome by imparting a flat or land to the cutting edge as shown in *Fig. 18* at 'B'. A narrow land is all that is needed, and this can readily be produced with a small slipstone. If much brass work is being undertaken in the workshop it is often found worthwhile to keep a number of drills, not necessarily a full set, for drilling brass only. In this way the inconvenience of having to resharpen a drill previously used on brass before it is again employed on steel can be avoided.

After much resharpening the chisel edge of a twist drill will be found to have materially increased in length. The reason for this is that the web of the drill is purposely made progressively thicker along the drill's axis in order to impart strength. Obviously a chisel edge that is too long will actively decrease the drill's free-cutting powers. These can, however, be restored by an operation known as thinning the point.

The thinning is brought about by grinding an equal amount from each side of the web until, at the drill point, it is approximately one-eighth of the drill's diameter. In professional establishments this is work normally given to operators supplied with special fixtures for the purpose. Nevertheless, an experienced workman will often do the job freehand. All that is needed is a suitably narrow grinding wheel and a steady hand, bearing in mind that little needs to be removed from the face of the web to achieve the desired object, see *Fig. 19*.

It will be appreciated that the angle of the chisel edge is not ideal since the chisel edge rubs rather than cuts. This rubbing tends to throw the drill off course resulting in a drilled hole out of axial alignment. An improvement to this condition can be brought about by modifying the point of the drill so that there are in effect two chisel edges as depicted in the illustration *Fig. 20*.

In this way the rubbing surface is reduced to nothing by the production of two new cutting edges that act as a pilot drill. This materially improves the rate at which the drill will cut. In order to carry out this modification it is really necessary to have available a

Fig. 19. Thinning the drill point

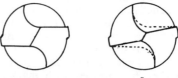

POINT TOO THICK   POINT THINNED

DRILLS AND DRILLING

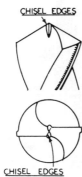

Fig. 20. Modified drill point

fixture that will allow the drill to be indexed accurately through 180 degrees. In addition a knife-edge wheel must be mounted on the spindle of the grinding head so that the wheel's periphery can reach into the flutes of the drill without the side of the wheel coming into contact with the drill itself.

The drill approaches the wheel in the manner and at the angle in the diagram *Fig. 21*. Readers, who for the most part will not be in possession of a suitable grinding fixture, should not be deterred from attempting the modification free-hand. If experiments are conducted with the larger drills, say $\frac{1}{4}$ in. diameter upwards, and these are the drills that benefit the most from the treatment, practice will soon make the worker, if not perfect, then adequate.

Drill Speeds

The speeds at which drills should be run are recommended by the manufacturers on the assumption that the drills are made from high-speed steel and are to be used in commercial production. In the amateur sphere there is no requirement to drive drills to the limit, so any figures given in commercial handbooks can be reduced with advantage.

The following table gives the drill manufacturers recommended speeds when drilling mild steel under commercial conditions. They can be reduced with advantage in the amateur shop.

Drill dia. in inches	High-speed drills r.p.m.	Carbon steel drills r.p.m.
$\frac{1}{16}$	4,000	1,800
$\frac{1}{8}$	2,000	900
$\frac{3}{16}$	1,500	600
$\frac{1}{4}$	1,100	450
$\frac{5}{16}$	900	340
$\frac{3}{8}$	750	280
$\frac{7}{16}$	650	240
$\frac{1}{2}$	550	210

When drilling brass and aluminium these speeds may be doubled, but should be materially reduced when cast iron is being drilled. If not the lands that form the sides of the drill may be worn away and the drill itself will then bind in the hole. From the table above, it will be apparent that drills made from high-speed steel are capable of withstanding a greater load than those made from carbon steel. Moreover, they retain their sharpness longer and are less likely to breakage than carbon steel drills.

Fig. 21. Grinding the modified point

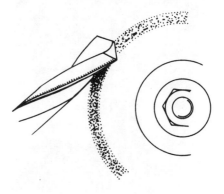

Drill Lubricants

Except when drilling cast iron or brass the work needs lubricating. At one time lard oil was much recommended for the purpose, but today there are many other lubricants available. For the most part these are thin-bodied so that they penetrate the work quickly. In the amateur workshop they are usually applied with a brush, but they can, if thought advisable, be used with a simple air-operated spray.

The use of mist lubrication has much to recommend it, particularly on machines where the lubricant, if used in any quantity, is liable to be splashed about.

An example of mist lubrication equipment, made originally to solve a problem in the sculptering of large aircraft components, is seen fitted to a pillar drill in the illustration *Fig. 21A*.

Cross Drilling

One of the more important operations the metal worker has to perform from time-to-time is the cross-drilling of shafts. This process is needed, for example, when a component such as a gear wheel needs to be fastened to the shaft. Unless, for some reason or other, the drilled hole must be offset it may be assumed that the hole has normally to pass through the centre line of the shaft.

When but a single shaft or component has to be cross-drilled the work is first marked off then mounted accurately under the drill point in a pair of V-blocks of the type illustrated in *Fig. 22*. These are supplied with one or more clamps enabling the work to be held firmly. The method of marking off and setting the work is depicted in the diagram *Fig. 23*.

Centre lines are first scribed upon the end of the shaft or component and one of the lines is extended along its axis. A radial line cutting the latter is then scribed and a centre marked with a punch at the intersection of the two lines that have been marked off.

When this part of the work has been completed a centre drill is placed in the chuck and the work aligned with it, using a square, as shown in the illustration, to ensure that the scribed centre line is truly upright. The work

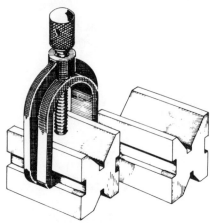

Fig. 22. A pair of V-blocks

Fig. 21A. Mist lubrication

DRILLS AND DRILLING

is now secured by clamping both in the V-block as well as on the table of the drilling machine, and drilling can commence using first the centre drill (1) followed by a twist drill of the correct size (2). If, for any reason, the specified drilled hole is large it is best to use a pilot drill before finally opening out the hole to size.

Frequently it is necessary to cross-drill a number of similar components. If the procedure just described had to be followed for every individual part it will be clear that the whole operation would be most time consuming not to say tedious. Industrially, of course, the matter is dealt with a jig or fixture made especially for the purpose. The amateur however, and even the professional engaged in small batch production, can solve the problem by means of a simple fixture that he can make for himself and which is illustrated in *Fig. 24*. The device consists of a V-block fitted with a pair of studs over which a saddle carrying a guide bush is placed. To ensure that the bush is correctly aligned before drilling commences it is only necessary to make sure that the distances between the underside of the saddle and the top of the V-block, marked by the letters 'A' and 'B' in the diagram, are equal.

The jig is fitted with a simple adjustable stop for use when a number of identical parts have to be cross-drilled. The parts of this unit are depicted in the illustration *Fig. 25*.

The top comprises a bar (5) passing through a clamp (2) and carrying a wishbone (4) fitted with a long and a short abutment; either of these may be brought into use and set at the correct distance from the work by turning the bar in the clamp and setting it where required. The stop is then secured with the lock screw (3). The details of these and all other relevant parts of the cross-drilling jig are given in the illustration *Fig. 26* and the

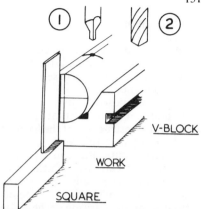

Fig. 23. Setting work for cross-drilling

Fig. 24. Cross drilling jig

Fig. 25. Parts of the jig adjustable stop

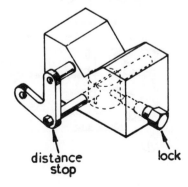

numbers quoted refer to this illustration.

The guide bushes, details of which are seen in *Fig. 27*, are made from mild steel so that they may be case-hardened if thought necessary.

It is not proposed to give a detailed description of methods that may be used in the construction of the fixture; for the most part the work is of an elementary nature. But one aspect of the procedure calls for a little comment. It will be readily appreciated that the success of the device depends entirely upon the position of the guide bushes in relation to the work seating in the V-block. If the bushes are in any way displaced off centre then the jig is useless.

In order to ensure that the bushes are correctly positioned it is essential that the clamp itself is accurately placed in relation to the V-block. This requirement can be met quite simply if the following procedure is adopted:

Referring to *Fig. 28*, the clamp (1) is first marked off, then drilled with two pilot holes, say $\frac{3}{32}$ in. dia. Next, a circular chamfered plug (2) is fitted to the clamp which is then secured to the V-block using a pair of toolmakers clamps for the purpose. The $\frac{3}{32}$ in. pilot drill is then fed into the V-block for $\frac{1}{2}$ in. and one end of the block and the corresponding end of the clamp are marked either with a centre punch or a letter punch so that the clamp and V-block can always be registered

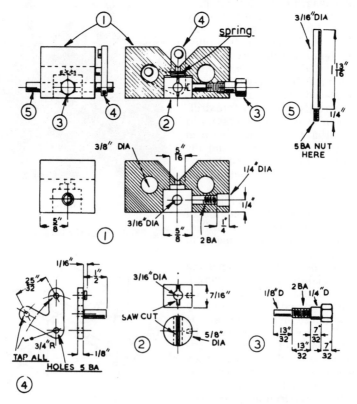

Fig. 26. Details of the drilling jig

DRILLS AND DRILLING

correctly. Only light marking, is needed at this stage to avoid possible displacement of the parts; final marking can be carried out later.

The pilot holes in the block are now opened out to the tapping size required. It is advisable to retain the clamp in place whilst this part of the work is being done; in this way the accuracy of the part's relationship will be maintained.

The holes in the clamp itself may now be opened out to clearing size over the studs fitted to the V-block. If the clamp remains in place during the process, and retained after this is completed, it may be used as a guide when the holes in the V-block are being tapped; but it is advisable to start the tapping in the drilling machine to ensure that the studs are truly square in the block.

One further class of cross-drilling jig must be mentioned. This consists of a block, fitted with a clamp, having pilot holes allowing such components as boring bars to be cross-drilled at any angle required. The fixture, when in use, is held in a machine vice, the vice, when the work is small, being allowed to 'float' freely on the table of the drilling machine.

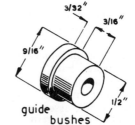

Fig. 27. Guide bushes for the jig

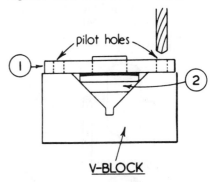

Fig. 28. Drilling the jig V-block

ods, that is to say by either counterboring, countersinking or spotfacing. The more important aspects of the subject are dealt with in the following chapter, Chapter 19.

Counterboring, Countersinking and Spotfacing

When bolts or screws are being fitted good practice dictates that they should be properly seated in or on the parts for which they are intended. In commercial practice, this is not always the case; the position, however, has been greatly improved by the introduction of die-casting, enabling, for the most part, accurately finished seatings for bolts and screws to be formed without recourse to any additional machining.

The amateur, however, will often be called on to form screw seatings by one or other of the recognised meth-

Positioning Mounting Holes on Machine Bases

When complete machine assemblies have to be mounted on wooden baseboards or metal frames it is sometimes difficult to establish by measurement the exact centres for the bolt holes that are needed.

In many instances it is possible to drill 'on assembly' as the saying is, but sometimes even this is not possible because of the difficulty of getting the drill to bear on the work.

It is then that a simple device we have used for many years will be found useful. This device, originally introduced to deal with wooden baseboards but equally applicable to metal framework when made from suitable steel, is the by-product of a parting-off operation in the lathe. As described in *Fig. 29* it consists of a piece of round material provided with a dimunitive centre-punch at one end.

The measurement 'D' is that of the hole in the machine assembly through which the device is passed in order that the centre punch can make contact with the baseboard *Fig. 30* at 'A' and produce a centre suitable for use with a machine wood bit, as illustrated in *Fig. 30* at 'B'.

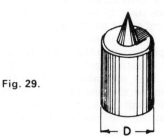

Fig. 29.

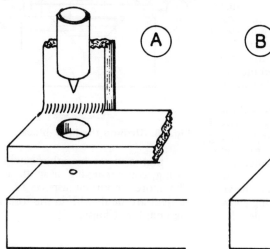
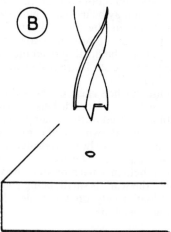

Fig. 30. Device for piloting holes in wooden base boards

CHAPTER 19

Countersinking and Counterboring

THE operations that give their names to the title of this chapter are frequently employed for a number of purposes in workshop practice. They are each depicted diagrammatically in *Fig. 1*, and, for the most part they are usually carried out in the drilling machine.

Countersinking

As will be seen in the illustration, countersinking is most commonly used to seat screws both in wood and metal. Provided the correct type of tool is employed the operation is one without difficulty, but it sometimes happens, either because the particular countersink is unsuitable, or the machine spindle lacks rigidity and is, perhaps, being run too fast, that vibration is set up resulting in a roughened finish to the screw seating as the pattern of the countersink itself is often reproduced on the work.

The countersink illustrated in *Fig. 2* is designed to combat this trouble. It is similar in form to the D-bit, described elsewhere in the book, and has also been produced commercially as may be seen in *Fig. 3* where a pair of single-edge countersinks, one of commercial origin the other home-made, are illustrated. These tools cut fairly fast and provide a perfect finish to a screw seating. They operate best if well lubricated and should be run at a moderate speed.

The three countersinks seen in *Fig. 4*, are standard commercially available cutters. 'A' is a 60 degree countersink for use on metal, 'B' is a 90 degree cutter of the 'snail' pattern, particularly useful when working in wood or plastics, whilst 'C' is a 90 degree tool of the type commonly employed when countersinking metal components.

Sharpness is an essential quality in cutters used for countersinking. Whilst, for the most part, badly blunted tools will need regrinding on

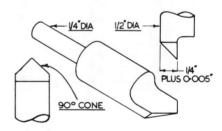

Fig. 2. Countersink with single cutting edge

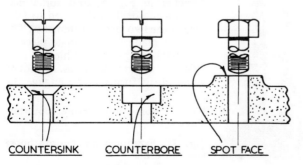

Fig. 1. Countersink, counterbore and spot face

Fig. 3. A pair of simple edge countersinks

special equipment, much may be done to improve their cutting edges by hand stoning in the manner outlined in the illustration *Fig. 5* taken from *Sharpening Small Tools* published by Argus Books Ltd. This depicts many of the different countersinks available and their application as well as the manner in which the hand stone should be applied.

Counterboring

When it becomes necessary to sink cheese-head or cap-head screws so that they are flush with the surface of the work the operation of counterboring is employed. As will be observed from the original illustration *Fig. 1* this involves the machining away from the original drilled hole a portion equal to the thickness of the screw head and of the same diameter as the head itself. When forming a counterbore three matters are important. The first, and perhaps most obvious one, is that the depth of the machining should be correct and should match the head thickness of the screw itself. The second requirement is that the diameter of the counterbore should fit the screw head closely or the appearance of the work will be spoilt. Finally both the hole and the counterbore must be in alignment or the screw may not enter. The cutter depicted in *Fig. 6*, and sometimes called a pin drill, is designed to take care of the two last requirements. The cutter is made from a single piece of tool steel machined and filed to provide a pair of cutting edges. A pilot or guide pin is furnished to engage the

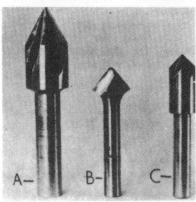

Fig. 4. Some commercial countersinks

Fig. 7. Examples of counterbores and pin drills

COUNTERSINKING AND COUNTERBORING

Fig. 6. A simple counterbore or pin drill

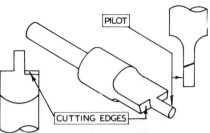

screw hole and keep the cutter in alignment. This pilot is turned from the parent material to a size that allows it to fit the screw hole closely.

Cutters of this type are readily made from silver steel, forming a useful medium when non-standard screws need to be accommodated.

Fig. 7 illustrates a number of counterbores and pin drills. It will be noticed that included amongst the collection is a commercial end mill. Used with care, after first partially forming the counterbore with a twist-drill, the end mill will perform satisfactorily. But it should be remembered that an end-mill also cuts on its side so any shake in the machine spindle may result in an oversize hole.

Twist drills themselves may be modified for use as counterbores. The method of doing so is illustrated in *Fig. 8*. First the point of the drill has to be removed by a grinding operation as depicted at 'A', then each lip in turn is backed off at an angle of 5 degrees as illustrated at 'B' where the drill itself is seen mounted in a V-block. One ad-

Fig. 5. Various types of countersink and methods of sharpening them

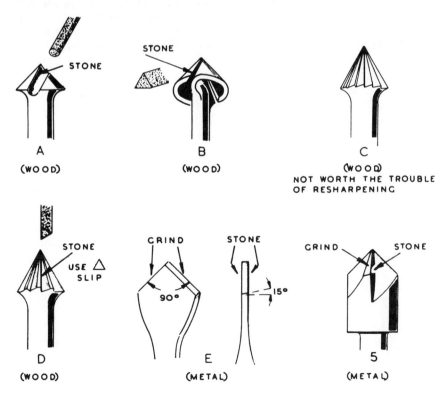

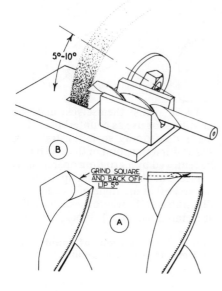

Fig. 8. Modifying the drill point

vantage of using twist drills for the purpose is that a wide range of counterbore sizes is readily available with the minimum of difficulty in making.

Spotfacing

Bolt or screw heads abutting against bosses on a casting need a machined surface that is smooth and is also square with the drilled hole through which the shanks of the bolts or screws are passed. Whilst a one-piece pin drill of the type already illustrated can be used to carry out the spotfacing operation in the drilling machine, it is common, particularly in large commercial undertakings, to provide sets of equipment where both the cutters and the pilots used are interchangeable.

In the small workshop, however, one-piece spot-face cutters are normally employed with one or more tools of the detachable cutter type as seen to the right of the group in *Fig. 7*.

CHAPTER 20 *Cutting Screw Threads*

IN the small workshop screw threads are normally formed by one of two methods. The first of these makes use of equipment held in the hand, while the second employs a machine, for the most part the lathe, to produce them. The amateur rarely needs hand operated equipment to produce threads above ½ in. in diameter since he can always resort to the lathe when he needs to cut threads of a larger size.

When discussing the whole subject of screw threads we are once again brought up against the problem of 'going metric'. In the years to come the newcomer may well find in the metric field a sufficient range to meet his requirements. At the moment of writing however, the amateur worker is well catered for by a number of different classes of screw thread. These are not going to disappear overnight and it is equally certain that the tackle for producing them will be available for a long time to come. For this reason the present position relating to screw threads forms the background to the notes that appear in this book, having in mind that in general methods and tools are similar despite minor differences in matters such as thread form and pitch.

Basic books providing information on aspects of the subject omitted here on account of space are available and some readers may wish to consult these.

So, on then, to the consideration of those classes of screw thread that are the most convenient for use in the small workshop. The purchase of suitable equipment will, for the most part, be a gradual process. Many readers no doubt already have a representative collection of tackle based on their immediate needs but the newcomer must decide for himself the class of the work he is to undertake at the outset and base his requirements upon this.

English standard screw threads are based on the Whitworth thread form introduced by Sir Joseph Whitworth over 100 years ago in an effort to standardise a screw thread that could be acceptable to all manufacturers instead of allowing each to fix his own standard.

For the most part, in this he was successful. But the pitch of the thread chosen for any particular diameter, that is the distance from one thread crest to the next, was such that the core diameter of small bolts and screws was proportionately small. As a result such components were much weakened and were liable to break in service. Consequently, industrialists, while wishful of retaining Whitworths thread form, decided that in relation to their diameter the thread pitches themselves needed some revision, their action culminating in the introduction of the British Standard Fine thread. As an indication of the practicability of this step a comparison of threads of the same form but different pitch on the same size of component is shown diagrammatically in *Fig. 1*.

The original British Standard Whitworth thread, however, still found an application, particularly amongst those concerned with castings in aluminium, where it was found that the coarser thread was superior when tapping holes for studs and the like; the greater area of metal in the thread itself much reducing any liability to

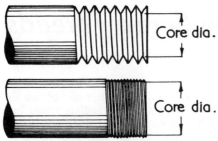

Fig. 1. The effect of coarse and fine pitch on core diameter

'stripping' in the comparatively soft metal.

Amongst the various classes of thread available for and suited to the amateur or the small workshop, based on the standard Whitworth form having an included thread angle of 55 degrees, are:

1. British Standard Whitworth, (B.S.W.) — where the pitch of the thread varies with the diameter
2. British Standard Fine (B.S.F.) — where the above also applies
3. Brass Thread — where all pitches are identical namely 26 threads to the inch.
4. Model Engineer — where all pitches are preferably identical, namely 40 threads per inch, though in some sizes a pitch of 26 threads per inch has been recommended.
5. British Standard Parallel Pipe (B.S.P.) — where the pitch of the thread again varies with the diameter.

In addition, and of particular use to the instrument maker is the British Association Standard (B.A.) This thread has an included angle of $47\frac{1}{2}$ degrees, the crests of the thread are more rounded than those of the Whitworth form, and the pitch varies with the diameter. It is a most convenient thread to use on much work below $\frac{1}{4}$ in. diameter.

Hand Screw Cutting Tackle

While at one time there were a number of differing pieces of equipment for producing male threads, today, for the smaller sizes at any rate, the button or circular die is most generally used. This form, split for the most part so that a degree of adjustment can be obtained, is held in a die-holder and is applied to the work by hand.

The elementary type of die holder has no form of guidance, so the starting and the maintaining of a thread square with the axis of the work is

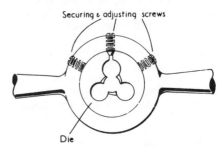

Fig. 3. Standard method of retaining die in holder

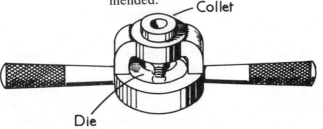

Fig. 2. Die holder fitted with guide collet

CUTTING SCREW THREADS

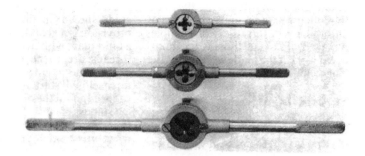

Fig. 4. Three colleted die holders

entirely dependent on the skill of the worker.

However, it is possible to make for oneself a form of die holder provided with guides that will enable the user to cut threads with the certainty that they will automatically be square with the component itself. This die holder is depicted diagrammatically in *Fig. 2* where the location of the guide and its relation to the circular die itself may be seen.

In the common die holder illustrated in *Fig. 3* the die is retained in place by three screws, one being used to expand the die when necessary. Such an arrangement by itself is well enough when no guides are used, but is of no use when they are. For this reason the guided or colleted die holders, three of which are illustrated in *Fig. 4,* have each a pair of axial screws designed to hold the die against its abutment face and thus square with the guide, as depicted in *Fig. 4A*. Three grub screws are employed, these serving to adjust the die for size, while the guide, as seen in the illustration *Fig. 5* is held in place by a thumbscrew. The body of the device is fenestrated to allow metal chips to be discharged.

Bodies for colleted die holders are readily made from mild steel shaped to the dimensions attached to the illustration *Fig. 6*. The measurements given cover the three sizes of die most commonly used, that is to say dies $\frac{13}{16}$ in., 1 in. and $1\frac{5}{16}$ in. diameter. The

Fig. 4B. The collets

Fig. 4A. The die seating and adjusting screw

Fig. 5. Die holder to show position of guide

handles can be machined to taste, suitable lengths and diameters being attached to the schedules.

Interchangeability of the dies and the collets or guides make this form of die holder very versatile. For example, it is possible to cut threads on shouldered work if the guide selected fits the work's major diameter.

Taps and Tap Wrenches

Modern taps, for the most part, follow the pattern depicted in the illustration *Fig. 7*. Few, however, are provided with the parallel lead or guide shown on the taper tap. This provision is a great help when starting a tap square with the hole in which the threads are to be cut; because, by drilling the hole to suit the lead, the tap itself will start upright and remain so throughout the operation.

Guides of this form are seldom found on taps below ¼ in. diameter so when using small taps and those not provided with a parallel lead, squareness of entry should be checked in the manner illustrated in *Fig. 8*.

Two classes of wrench are commonly employed. The first, depicted in *Figs. 9* and *9A* for use with taps up to ¼ in. diameter and usually available in two sizes; the second, suitable for larger taps and obtainable in a variety of forms but generally following the pattern seen in the illustration *Fig. 10*.

Equipment for Threading in the Lathe

Threads may be cut in the lathe in either one of two ways. The first of

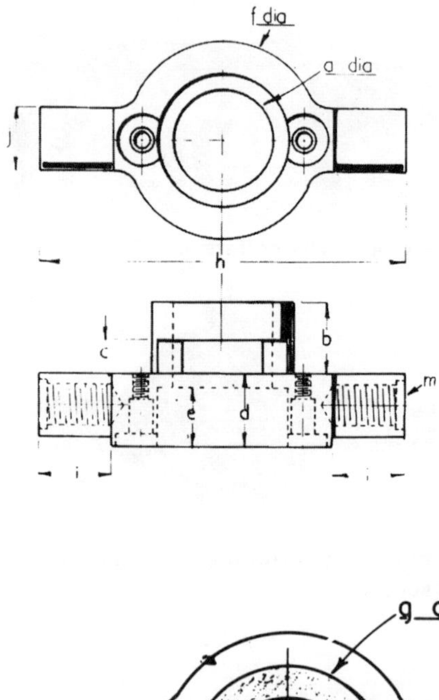

Fig. 6. Details of die holder body

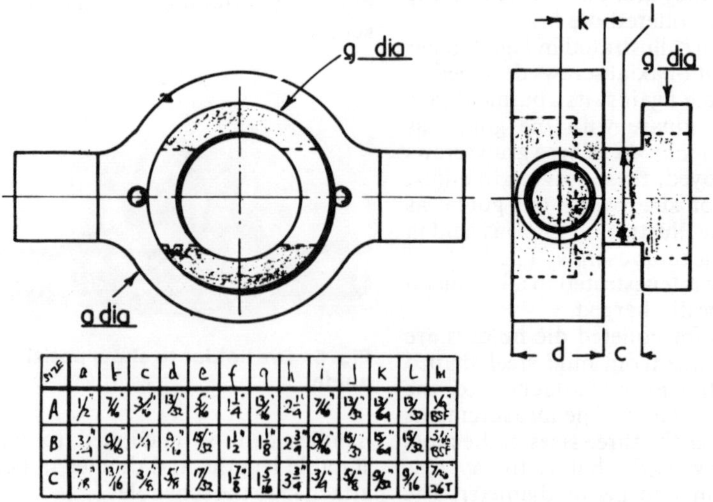

Fig. 6A. Details of die holder body

SIZE	a	b	c	d	e	f	g	h	i	j	k	l	m
A	1/2"	7/16"	3/16"	13/32"	5/16"	1 1/4"	13/16"	2 1/4"	7/16"	13/32"	13/64"	13/32	1/4" BSF
B	3/4"	9/16"	1/4"	9/16"	15/32"	1 1/2"	1 1/8"	2 3/4"	9/16"	15/32"	15/64"	15/32	5/16" BSF
C	7/8"	13/16"	3/8"	5/8"	17/32	1 7/8"	1 5/16	3 3/4"	3/4"	5/8"	9/32"	9/16"	7/16" 26T

CUTTING SCREW THREADS

Sc..p	a	b	c	d	e
A	2⅞"	7/16	1⅛"	¼ BSF	5/16"
B	3½"	9/16	1⅜"	5/16 BSF	⅜"
C	5½"	11/16	2"	7/16 26T	½"

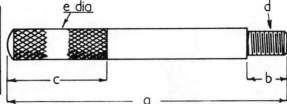

Fig. 6B. Details of die holder handles

these makes use of hand equipment suitably mounted, for the most part in the tailstock, to ensure that the screw thread is in axial alignment with the work. Most lathe manufacturers can supply die holders, such as that seen in *Fig. 10A*, that are supported on a peg held in the tailstock and that will travel along it as the thread is formed, while small taps may be gripped in the tailstock drill chuck.

The colleted die holders previously described may be used in this way if the handles are shortened and one is provided with a tubular extension to engage the lathe top slide, so preventing the device from turning. The parts of the equipment are illustrated in *Fig. 11* whilst the die-holder is seen in operation in *Fig. 12*.

When the larger size of tap is mounted in the tailstock chuck it is usually necessary to make some provision in order to stop the tap and chuck turning. Perhaps the simplest way of ensuring that they do not do is to grip the tap with a lathe carrier in the manner indicated by the illustration *Fig. 13*. The carrier can then be brought into contact with the top slide.

For those who would prefer a somewhat more sophisticated device, suitable for the smaller size of tap, the tool illustrated in *Fig. 14* and *Fig. 14A* may appeal. It has been designed to accommodate taps up to ¼ in. shank diamete and consists of a short length of hexagon material 'A', cross-drilled to accept two short arms 'B' and provided with a set screw 'D' to grip the tap. In common with the die-holder previously illustrated this tap holder has

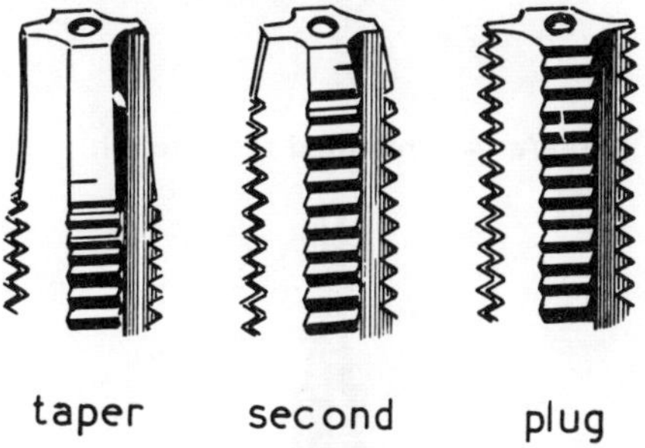

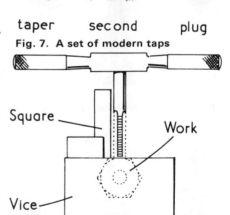

taper   second   plug

Fig. 7. A set of modern taps

Fig. 8. Checking squareness of tap

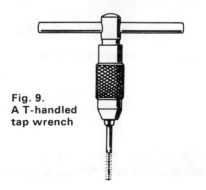

Fig. 9. A T-handled tap wrench

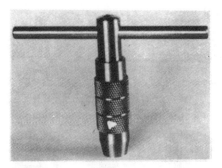

Fig. 9A. A T-handled tap wrench

Fig. 10A. The Myford die holder

Fig. 11. The tailstock die holder

an extension arm 'C' that may be brought into use when the top slide of the lathe is needed as a stop.

Holding and starting really large taps in the lathe often poses a problem, not always easy to solve. Two solutions to the difficulty, however, are illustrated in *Fig. 15* and *Fig. 16* respectively. In the first example the tap is located by the tailstock, mounted on the back centre, and is turned by a wrench while being fed forward by the tailstock hand wheel. Once the tap has properly engaged it will feed itself, but care must, of course be taken to follow the tap or it will probably run out of alignment.

The second solution provides for the tap to be gripped in the toolpost on the top slide and aligned with the work by means of packing. The tap is fed into the work by means of the saddle which is free to move along the lathe bed so as to provide no constraint. In both instances the work can be transferred to the vice for completion of the tapping by hand. Tapping may also take place from the headstock. The process in its simplest form the tap is held in the chuck while the work is supported by the tailstock drilling pad and secured by toolmakers clamps. This arrangement is depicted diagrammatically in *Fig. 17* though, for more clarity, the clamps have been omitted from the illustration. The method is generally suitable for plate work.

A second procedure makes use of the top slide and tool clamp to hold the work, and a simple example of the manner in which it is used is illustrated

Fig. 10. Wrench for large taps

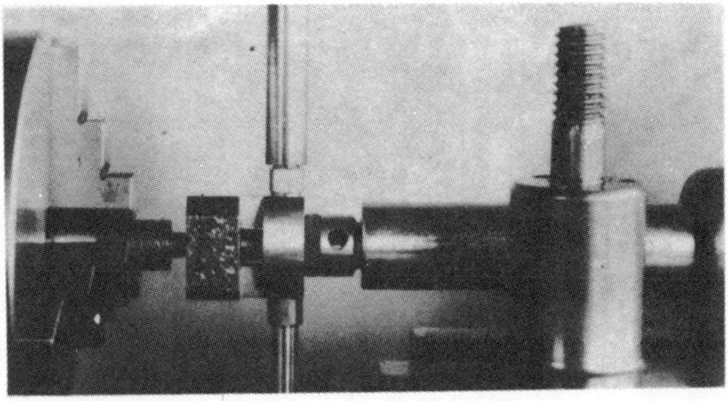

CUTTING SCREW THREADS

Fig. 12. The tailstock die holder in use

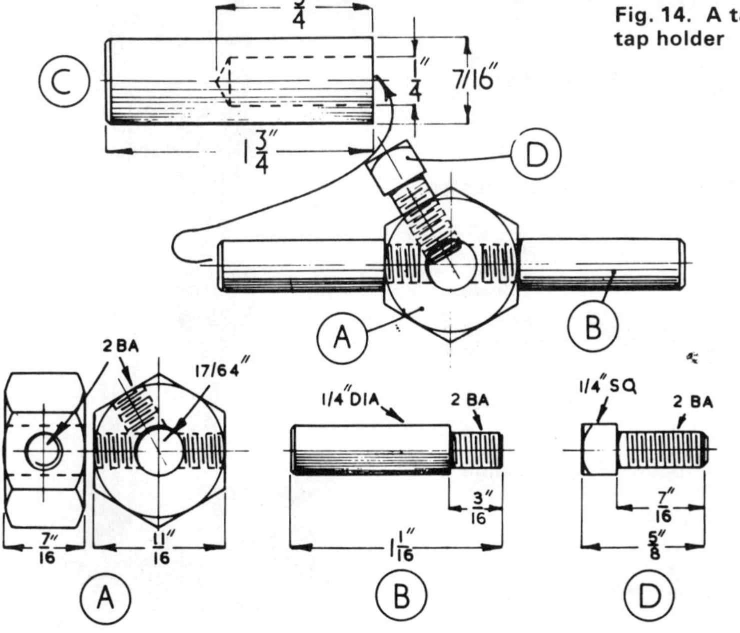

Fig. 14. A tailstock tap holder

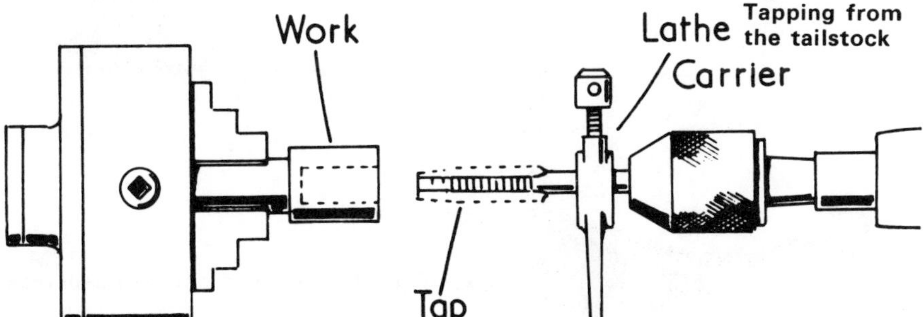

Fig. 13. Tapping from the tailstock

Work

Lathe Carrier

Tap

Fig. 14A. A tailstock tap holder with extension handle

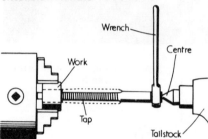

Fig. 15. Tap mounted on back centre

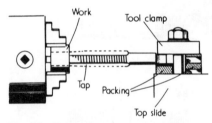

Fig. 16. Tap mounted on top slide

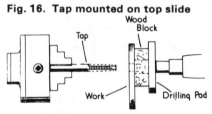

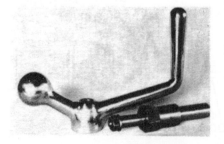

Fig. 23. A drilling machine converted for tapping

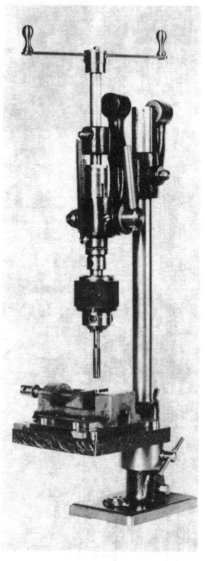

Fig. 17. Tapping from the headstock

Fig. 20. Myford mandrel handle

CUTTING SCREW THREADS

Fig. 19. Drummond mandrel handle

Fig. 20A. Another form for Myford mandrel

Fig. 18. Starting a large tap from the headstock

in *Fig. 18*. The work is first set square with the chuck and at the correct height so that it can be machined with a flycutter as depicted at 'A'. Next the work is prepared for tapping with a series of drills as seen at 'B'. The purpose of the counter drill is to remove metal from the area of the first two threads for a reason we shall presently explain. A taper tap is then gripped in the chuck and fed into the work until

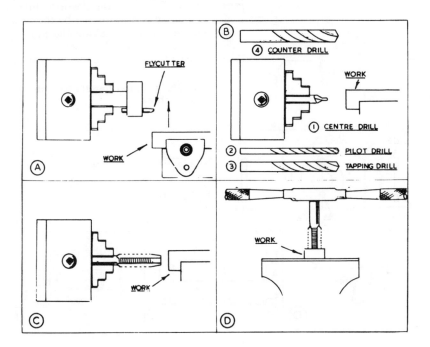

it is well started as shown at 'C'. When this has been done the work can be removed and placed in the bench vice for hand tapping as depicted in the diagram 'D'.

Mandrel Handles

When tapping in the lathe it is generally inconvenient, if not actually detrimental, to drive the mandrel under power, so much advantage may be gained by fitting some form of handle that will enable the mandrel to be turned with a sensitivity that would otherwise be impossible. Some ingenuity will need to be exercised in this matter because accessibility will play an important part in the eventual arrangement adopted. Two examples are handles that were made and fitted by us to the Drummond and Myford lathes respectively. As will be noted the Drummond handle screws directly into the hollow mandrel, which was tapped out to accept it, whilst that fitted to the Myford is secured to the adapter made for the change wheels used for dividing purposes and described in Chapter VII. The handle itself is a diecasting, but a built-up substitute will do equally well. The two handles described are illustrated in *Figs. 19* and *20*.

Tapping in the Drilling Machine

Undoubtedly the drilling machine is a very convenient medium for starting a tap and even for carrying out the complete tapping operation. Fundamentally, the machine is used in the way demonstrated by the diagram *Fig. 21* where it will be seen that a lathe carrier is attached to the machine's spindle to act as a handle, whilst the work itself is placed on the table. The driving belts are removed, of course, in order to maintain the sensitivity of the set up. For those who contemplate much work of this nature the drilling machine handle illustrated in *Fig. 22* may be found worthwhile. It is secured to the spindle by the hexagon screw 'D' engaging the keyway. The handle may be extended as required by sliding the bar 'E' through

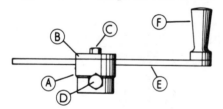

Fig. 22. Tapping handle for the drilling machine

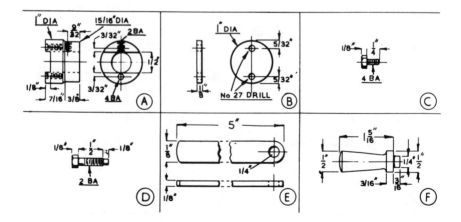

CUTTING SCREW THREADS

the body 'A' and securing if by the clamp 'B'. Some may prefer to fit a key to the body instead of relying on the set screw to take the drive. If this is so the key may easily be made fast with Araldite, thus avoiding the difficulty of securing with small screws.

If thought desirable the handle may be extended and fitted with a pair of knobs as depicted in the illustration *Fig. 23* which shows an old drilling machine that has been adapted for tapping purposes and since converted to power operation.

In order to obtain the best work the vice needs to be so mounted that is is free to align itself with the tap. This universality is achieved by attaching the vice to a sole plate working as a slide and superimposed on another slide at right angles to it, the whole being bolted to the table of the drilling machine. The arrangement being known, for obvious reasons, as a 'jelly plate' mounting.

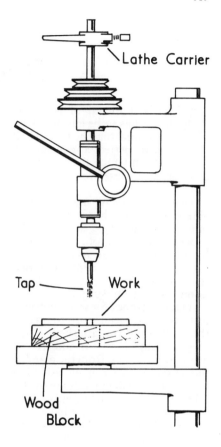

Fig. 21. Tapping in the drilling machine

CHAPTER 21

Cutting Screw Threads in the Lathe

WHEN a thread of large diameter has to be cut or when it is essential that the thread itself should be in perfect alignment with the part upon which it is to be cut, then the lathe is used to carry out the work.

There are primary manuals available most of which provide a basic outline of the lathe screw cutting process, but no apology is offered for recapulating the basic information in part, since the intention here is eventually to extend the readers knowledge of the subject.

A lathe intended for screw cutting is provided with a lead screw to traverse the saddle along the lathe bed. A train of gears can be set up and used to drive the lead screw from the mandrel, the ratio of the gearing dictating how far the saddle will move along the bed for each turn of the mandrel.

If now a tool is mounted on the saddle and engaged with work attached to the mandrel, as soon as the latter is turned a rough screw will begin to be formed, the pitch of this screw being dependent upon the ratio of the gearing employed.

Screw Cutting Tools

The basic screw cutting tools are illustrated in *Fig. 1*, the first intended for threading external work whilst the second is employed in internal screw-cutting. The angle of the point in both instances is ground to conform with the standard it is desired to cut. For example, if the Whitworth standard was being used then the included angle would be made 55 degrees.

Threads produced by these simple tools have crests and roots that are sharp, but, as an examination of any commercially-produced and well made bolt or screw will show, the practical thread has these details in rounded form and it is the function of the chasers depicted in *Fig. 2* to provide this finish to the work. These tools are sometimes applied by hand, but today it is preferred to set the chasers in the toolpost where they can be used under controlled conditions.

Of course, in industry, form cutters are used to produce the same results. These tools are, in effect, very short lengths of chaser, capable of cutting fully formed threads without a change of tool. For those readers who have good contacts in industry it may be possible to obtain old or unwanted chasers from the geometric dies used in the capstan lathe section of the works. Provided a suitable holder is made for them these old chasers often make very good form cutters.

Fig. 1. The basic screw-cutting tools

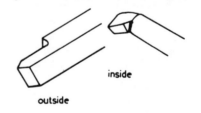

outside   inside

Fig. 2. Inside and outside chasers

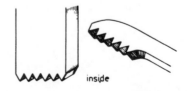

inside

Fig. 3. A height gauge for tools

When grinding a tool for screw-cutting, the sides of the point must be given sufficient clearance to prevent the tool from rubbing on the work. The extent of the clearance required will depend upon the helix angle of the thread in relation to its axis and is dependent upon the pitch of the thread itself; for work of a given diameter the greater the pitch the greater the helix angle. This angle can be found by the graphical method depicted in the diagram *Fig. 4*. Here the circumference of the thread

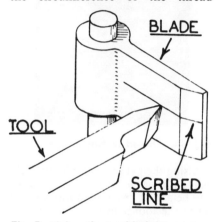

Fig. 5. Using the tool height gauge

Readers should note that it is essential all tools used for screw cutting are set on the lathe centre line, or an accurate thread will not be formed. The height setting gauge illustrated in *Fig. 3* will be found a great help in ensuring that they are set correctly. Little need be said about its construction for the simplicity of the device will be evident from the illustration. Not only will this gauge prove useful when screwcutting but it is also essential for setting the ordinary turning tool, particularly when tapers are being machined for this is another occasion when incorrect height setting will result in inaccuracy.

Fig. 4. Graphical method of obtaining the helix angle

$(3 \cdot 1416 \times$ the major diameter) is set out as a base line whilst the pitch is drawn as a line at right angles to it. If the triangle is completed the helix angle of the thread can be measured directly with a protractor. It follows then that the clearance angle needed at the leading edge of the tool must be greater than the helix angle of the thread to be cut. This is particularly the case when square threads are being machined, as we shall presently see. When making the height setting gauge a line is scribed and then engraved on the blade of the device making use of a centre placed in the tailstock to perform the necessary marking off.

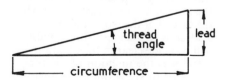

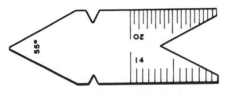

Fig. 6. Thread setting gauge

Fig. 6A. Thread setting gauge

Using the gauge is simplicity itself. All that is needed is to pack or adjust the height of the tool point so that when the tool is firmly clamped, the point is aligned with the scribed line on the gauge blade

The method is depicted in the diagram *Fig. 5*.

Fig. 7. Using the thread setting gauge

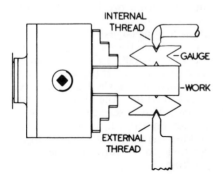

In addition to setting the tool at the correct height, it must also be aligned square with the work itself. This is the purpose of the gauge that appears in the illustration *Fig. 6*. It is, of course, a production of fine tool manufacturers and is used for both external and internal threading, being applied to the tool and the work in the way demonstrated by *Fig. 7*. Gauges of this type are available, for the most part, to suit various standards, the notches having included angles to match the particular thread to be cut. Thus, if Whitworth is the standard in question, the notches will have an included angle of 55 degrees. This enables the operator to grind the screwcutting tool correctly by checking its angularity against the large notch at the end of the gauge.

The Thread Indicator

At the beginning of the chapter reference has been made to the leadscrew. This is provided with a clasp nut, a device attached to the saddle enabling the leadscrew to be thrown in or out of engagement at will. It consists of a pair of half-nuts and a mechanism by which they may be caused to close upon or be released from the screw at will. For the most part the light lathes supplied for amateur use have leadscrews of $\frac{1}{8}$ in. pitch, or 8 threads to the inch. If a thread of this pitch was to be cut the leadscrew could be engaged wherever the clasp nut would drop in because the threads are the same. The same holds true for all sub-multiples of $\frac{1}{8}$ such as $\frac{1}{16}$, or $\frac{1}{32}$.

On the other hand, supposing a thread of $\frac{1}{12}$ in. pitch is to be cut, then the nut cannot be closed except when the thread of the work and that of the leadscrew coincide. This condition will be made clear in the diagram *Fig. 8* and it is the function of the thread indicator to show the lathe operator when this coincidence occurs so that

CUTTING SCREW THREADS IN THE LATHE

he can close the clasp nut at the right point.

A typical indicator is illustrated in *Fig. 9*, whilst its general construction is shown in *Fig. 10*. The indicator shown, consisting of three major parts, was made for the Drummond lathe. A bracket 'A' is used to attach the device to the saddle, whilst a swivel 'B' allows the indicator body 'C' to be moved so that the pinion mounted on the end of the indicator shaft can be brought into contact with the lead screw and then locked in place.

The pinion itself has 16 teeth meshing with the $\frac{1}{8}$ in. pitch lead screw. This causes the head of the indicator to revolve, the engraved numbers registering in succession with the zero index line on the body. The intervals between the figures quartering the dial each represent four turns of the leadscrew or $\frac{1}{2}$ in. of saddle movement, so the figures themselves can be used for all even-numbered pitches. Opposite pairs of figures, 1 and 3 and 2 and 4, represent 8 turns of the leadscrew, or a saddle travel of 1 in., so serving for the cutting of odd numbered thread pitches. When the claspnut is closed at any one number only, the distance travelled by the saddle is 2 in. so this particular number can be used as an indicator for half-pitches such as $12\frac{1}{2}$ threads to the inch.

Change Wheel Gearing

We have already seen that it is the relative speeds of the mandrel and the leadscrew that governs the pitch of the thread to be cut. For example, if the leadscrew has eight threads to the inch and the mandrel, and the work, turns at twice the speed of the leadscrew, then the pitch of the thread cut will be $\frac{1}{16}$ in.

It is the purpose of the change wheels to provide the gearing necessary to satisfy the ratio between the

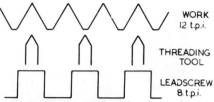

Fig. 8. Relation of tool point to leadscrew pitch

Fig. 9. The thread indicator

Fig. 10. Construction of the thread indicator

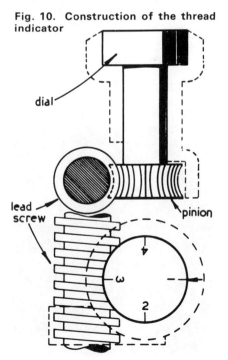

Fig. 10A. Parts of the thread indicator

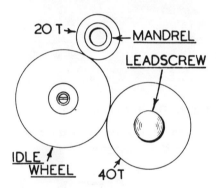

Fig. 11. A simple gear train

Fig. 12. Tumbler gearing

pitch of the leadscrew and that of the thread to be cut. If the cutting of a thread of $\frac{1}{16}$ in. pitch, or 16 threads to the inch, is taken as an example then the gears to produce this pitch using a lead screw of $\frac{1}{8}$ in. pitch can be calculated from the ratio:

$$\frac{8 \text{ (leadscrew thread)}}{16 \text{ (thread to be cut)}}$$

If gear wheels having 8 and 16 teeth were available these would be satisfactory but wheels having 40 teeth and 80 teeth, or 20 teeth and 40 teeth, would provide the same ratio; and as these are among the set of change wheels normally supplied with the lathe they could be used. It will, of course, not be possible to mesh the wheels directly so an intermediate wheel must be engaged with them. However, this does not affect the gear ratio in any way; moreover, the interposing of a third, or idle wheel as it is termed, is needed to ensure the production of a right-handed thread, that is unless the lathe is provided with tumbler gearing that allows the direction of the leadscrew to be reversed at will.

The location of the gears themselves is shown by the rule:

$$\frac{\text{Numerator}}{\text{Denominator}} = \frac{\text{Driving Gear}}{\text{Driven Gear}}$$

If the change wheels necessary to cut a thread of $\frac{1}{16}$ in. pitch are again

CUTTING SCREW THREADS IN THE LATHE

taken as an example the 20 tooth gear is mounted on the mandrel whilst the 40 tooth wheel is placed on the leadscrew. The diagram *Fig. 11* indicates the disposition of these gears.

The idle wheel is mounted on a stud attached to the change wheel bracket and from this location the mounting is adjusted to bring the wheel into mesh with the other two gears.

When the lathe is provided with tumbler gears the mandrel gear itself is not actually mounted on the mandrel but on a stud wheel attached to the tumbler gear bracket. Intermediate wheels, of course, have to be placed between the mandrel and the stud wheel but these do not, in effect, alter the gear ratio because the wheel permanently attached to the mandrel and the stud wheel itself are both the same size. The arrangement is illustrated in *Fig. 12A*.

The stud wheel has an extension fitted with a key to which the selected change wheel is attached.

Compound Gearing

For the most part, the numbers of change wheels supplied with a lathe are limited, so it is not always possible to employ a simple train of gears. In this event a compound train of wheels has to be used, but the precise details seldom need calculation by the operator because lathe manufacturers almost always provide a chart showing the change wheels needed for a wide range of threads.

However, it may sometimes be necessary to find the wheels to cut a thread not included in the lathe makers list, so it is as well to know how to set about the calculation. Let us take an example a thread of 15 turns to the inch. Now the ratio of the wheels to satisfy this pitch using an 8 to the inch leadscrew is 8/15 or 16/30. If we had these wheels they could be used directly, but, usually only the 30

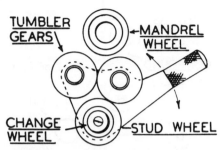

Fig. 12A. Tumbler gearing

tooth wheel is available so this is attached to the leadscrew and the 16 value in the ratio obtained by connecting a 20 tooth wheel on the mandrel stud to compound gearing. A 20 wheel has a common factor of 4 with the value 16 so, as 16 is four-fifths of 20

$$\frac{16}{20} = \frac{4}{5}$$

the compound gearing consists of two change wheels mounted on the same stud and coupled together, one having 20 teeth the other 25. If, because of availability, it is more convenient the compound gears can be 40 and 50 teeth, it will not affect the ratio.

We now have found all the necessary change wheels and these are connected as shown in the diagram *Fig. 13* But before doing so, however, the train of gears should be checked to ensure that they are indeed correct for the particular thread it is desired to cut.

Proving the Gear Train

The gears are proved in the following manner:

Threads per inch to be cut =
Driven gears multiplied together
over
Driving gears multiplied together
multiplied by Pitch of leadscrew expressed as t.p.i.

If we again take as our example the wheels to cut 15 t.p.i.

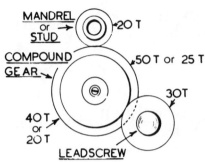

Fig. 13. Compound gear train

Fig. 14. Thread pitch gauge

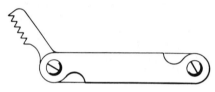

$$\frac{\text{Driven gears } 25 \times 30}{\text{Driving gears } 20 \times 20} \times 8 \text{ t.p.i.} = \frac{7 \cdot 5}{4} \times 8$$

$$= \frac{60 \cdot 0}{4} = 15$$

Except to check that the wheels they recommend are set in the right positions there is little point in proving the lathe manufacturers gear trains. These will have been checked many times in their drawing offices. An occasional exercise, however, on the lines we have demonstrated and the comparing of the results with the manufacturers solution will often prove useful as a refresher.

Practical Screw Cutting

We have seen the tools used in screw-cutting and have also discussed the way in which the lathe is set up to machine the pitch of thread desired, so the time has now come to examine the practical aspects of the work.

A screw thread has two diameters. The first the top or major diameter, the second the core or minor diameter, and we are interested in them because, by subtracting one from the other and dividing the result by 2, we are able to obtain the theoretical depth of the thread and thus the amount of in-feed that must be given to the tool itself.

For the most part the whole depth of thread, for any given pitch, is obtainable from books of reference. If faced, therefore, with a component for which a mating thread has to be cut, the pitch must first be ascertained. This is the purpose of the thread gauge illustrated in *Fig. 14*. It contains a number of leaves each having a different pitch cut upon it. By applying successive blades to the part the pitch of the thread can be determined. Gauges of this type are obtainable in both Whitworth and metric pitches, and those who contemplate much work of this nature would do well to add them to their tool kit.

Before the tools for screwcutting are actually set up, and after the work has been machined to the correct size, it is advisable to turn an undercut or run-out where the thread finishes, at the same time the lead end of the work can be chamferred or rounded to taste.

Some operators, after a light cut has first been taken with the threading tool, allow the thread to run out into a hole drilled in the work. This method is illustrated in *Fig. 15*, at 'A'. For the inexperienced worker, however, the undercut depicted in the same illustration at 'B', offers many advantages. This form of run out is turned with a small parting tool fed into the work for a depth equal to that of the thread to be cut. Not only does this provide some latitude when throwing out the clasp nut and disconnecting the lead-screw at the end of each cut, but when the point of the tool touches the surfaces of the undercut the worker knows that the thread has been machined to full depth.

In practice the width of an undercut

varies to suit the work in hand, the narrower it is the neater the work. However, until practice has perfected the operator's technique it is suggested that a minimum width of $\frac{1}{16}$ in. be used.

Measurement of the diameter is undertaken with a pair of outside callipers having their jaws fined down until they are some 0·030 in. thick at the extremity. The callipers are then set against packing of the right thickness, a drill of the correct diameter or a pair of inside callipers adjusted with reference to an outside micrometer.

At this point it may not be out of place to consider the three types of calliper used in thread measurement. If *Fig. 16* is examined it will be seen that the callipers used for measuring the outside diameter of male threads have wide spatulate extremities enabling them to bridge a number of threads thus providing a reliable reading. The inside callipers depicted at 'B' have pointed ends enabling them to measure the major diameter of internal threads, whilst the callipers at 'C' are the type that have already been referred to in connection with undercuts and run outs. For the most part measurements taken with these instruments are referred to as micrometers or verniers so that numerical readings can be established.

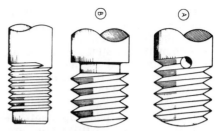

Fig. 15. Thread run-cuts

Cutting the Thread

There are two ways of feeding the tool into the work. In the first the tool is moved inwards by the cross slide, the direction being at right angles to the axis of the work, with the in-feed measured by the cross-slide index. The second method makes use of a technique that, used with care, effectively avoids the possibility of the tool digging in. Here, the top slide, sometimes called in American text books the 'Compound Slide', is set over at an angle. The obliquity is that of half the included angle of the thread to be cut.

Thus, when cutting threads to the Whitworth standard, the top slide is set over $27\frac{1}{2}$ degrees. The object is to make the leading edge of the tool do the bulk of the cutting whilst the trailing edge only shaves the work. The

Fig. 16. Thread callipers

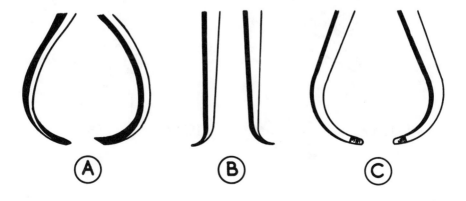

set-up can be used for both external and internal threads in the way depicted by the diagram *Fig. 17*. The cut is put on by the top slide feed screw, its index being used to measure the amount of in-feed. In this connection it should be noted that when the slide is set over in this way a reading of 0·001 in. on the index represents an actual movement of 0·0008 in. approximately, to be precise 0·00081 in.

The feed screw of the cross slide is only used to move the threading tool out of contact with the work at the end of each cut, and has its index set at zero, while the tool point is just touching this surface of the work. In this way, each time the cross-slide is withdrawn past the zero mark on the index the tool will clear the work effectively.

Before commencing the screw-cutting operation, however, the top slide index must also have been set to zero. This must have been done at the same time as the cross slide index was zeroed with the tool in contact with the work. But it will also be necessary to engage the thread indicator. If this has a movable index ring, adjustment is simple. However many indicators have no adjustment, they must therefore be set so that one of the engraved figures on the dial registers with the index line **before** the slide indices are adjusted in the manner described above, and, of course, with the lead-screw clasp-nut closed.

The saddle can now be moved to the right of the work, the leadscrew engaged with reference to the thread indicator, a cut of some 0·002 in. put on by the top slide, the back gear engaged and a trial cut taken. It is suggested that, when only a fine thread has to be cut, the lathe is best turned by hand leaving power operation for the coarser threads needing the removal of much material. Suitable equipment for turning the lathe by hand has already been described earlier in the chapter.

When the first cut has been completed, the lathe is stopped, the lead-

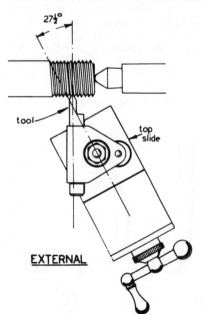
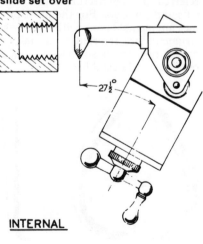

Fig. 17. Turning threads with top slide set over

screw disengaged, the tool withdrawn from the work and moved back to the beginning again so that the whole operation can be repeated; this process is continued until the thread is fully formed.

At first, and before attempting to screw thread a finished piece of work, the newcomer would be well advised to practise on a piece of material of no importance. In this way he can perfect his technique. When sufficient experience has been obtained he will be able to tackle a screw-cutting job with confidence.

The work needs plenty of lubricant. Lard oil, or one of the many specially prepared oils that are obtainable can be applied with a brush or via the suds system about which we shall have something to say later.

As to the depth of cut; until experience is gained light cuts are safest. But one should remember that the deeper the machining extends, the greater is the work surface in contact with the tool. Therefore the operator may well start with a relatively deep in-feed, say from $0 \cdot 005$ in. to $0 \cdot 007$ in. finishing off with lighter cuts of from $0 \cdot 002$ in. to $0 \cdot 001$ in. as he nears the end of the threading operation.

CHAPTER 22   *Measuring Equipment*

Most reference books covering workshop practice for beginners include something on the subject of equipment used to measure work in progress, but frequently this is limited to basic measuring tools such ruler, callipers, protractor, almost certainly the micrometer and possibly the depth gauge. These are the fundamentals and can be considered virtually essential requirements for the newcomer. Such books normally describe the use and care of such tools, and it is therefore taken the the reader has some acquaintance with them.

For this reason it is not proposed to spend further time describing the micrometer calliper, for example, as one must assume that readers will already have experience of the tool and know how to use it. Rather, space will be given to the use of the dial test indicator and its use, as well as the employment of other aids to accurate measurement and the setting of work for machining.

In the succeeding chapter, on 'marking off', full coverage is given to the tools needed. Many of these come into the category of measuring equipment but it is thought better to deal with them as they are applied to the specific work of marking off rather than describe them in the catalogue sense.

The Micrometer Fig. 1

As has already been assumed readers will be aware of this measuring device and how to use it. They will also doubtless know that micrometers are available to measure both inside and outside work. The inside micrometer, however, is a somewhat unsatisfactory tool, not capable of measuring the smaller holes at all, and, unless continuously handled, liable to provide a series of readings of problematical accuracy.

For the purposes of the amateur and small workshop it is far better to make use of the familiar outside micrometer in conjunction with particular devices that enable holes as small as $\frac{1}{8}$ in. to be measured with great precision.

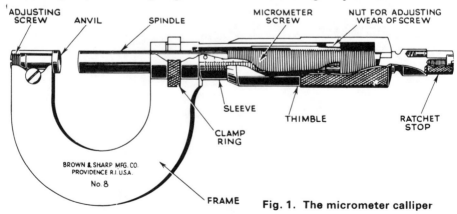

Fig. 1. The micrometer calliper

MEASURING EQUIPMENT

Fig. 1A. Micrometer stand

Micrometer Stands

When using a micrometer to measure a number of small objects it is usually more convenient to mount the micrometer in a stand leaving the hands free to hold the work and make the measurement.

The stand shown in the illustration *Fig. 1A* is a commercial product; but the making of a simple stand is work that can be undertaken in the small shop and will form a useful and instructive exercise.

Readers may well be able to design such a stand for themselves; for those however, who may not be quite clear as to the requirements of the device it is no more than a padded clamp with angle adjustment on a weighted base.

Small Hole and Telescopic Gauges

The most useful of the devices that can be used with the micrometer are the small hole and telescopic gauges.

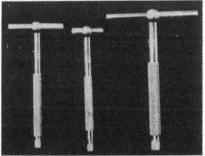

Fig. 2B. The telescopic gauges

The small hole gauge, illustrated in *Fig. 2A* consists of a steel ball machined integrally with a stalk enabling it to be mounted in a tubular handle. The ball and its stalk are split and the two halves are seperated by a movable wedge controlled by an adjusting screw situated at the base of the handle.

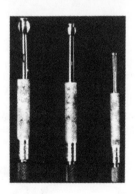

These small hole gauges are usually made in sets of four with a range from

Fig. 2A. The small hole gauge

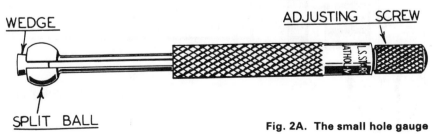

Fig. 2A. The small hole gauge

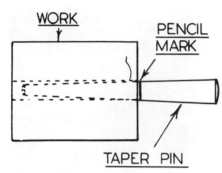

Fig. 3. Using a taper pin to measure a small hole

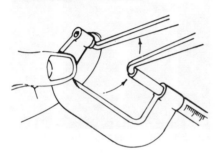

Fig. 4. Measuring callipers

⅛ in. to ½ in. They have the great advantage that they enable the hole's parallelism or otherwise to be checked immediately.

When the internal dimensions of the work are above ½ in. diameter the telescopic gauge, depicted in the illustration *Fig. 2b* provides a ready means of measuring a bore accurately.

The plunger, sliding in a fixed member attached to the handle, is spring loaded and can be locked by the finger screw passing through the handle. To use the device, the locking screw is first released, the measuring head is then passed into the work and the plunger locked taking care to see that the handle is being held horizontal. The gauge is then withdrawn and has an outside micrometer applied to its measuring head in the same way as the small hole gauge previously described.

Both the tools described are, of course, for the more advanced worker, so it may well be the case that more elementary methods will have to be adopted by the newcomer to workshop practice.

If it is accepted that their bores are parallel, small holes can readily be measured with a standard taper pin applied as shown diagrammatically in *Fig. 3.*

The taper pin is pushed into the hole to be measured and a pencil mark made on the pin adjacent to the edge of the hole. A micrometer measurement taken over the pencil mark then gives the size of the hole for all practical purposes. The tapered pin method is applicable to holes up to

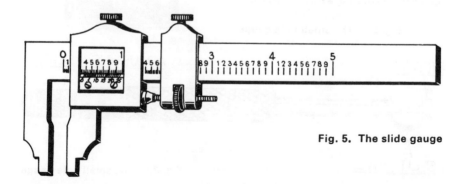

Fig. 5. The slide gauge

MEASURING EQUIPMENT

Fig. 6. The vernier

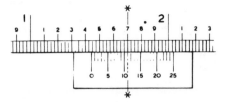

½ in. in diameter, though it may be necessary to make for oneself pins to suit the larger sizes. Above ½ in. inside callipers will need to be used, applying an outside micrometer to the jaws. This is a technique that needs practice to get the feel of the callipers both in the work and between the measuring surfaces of the micrometer. The method used is to hold the micrometer in the left hand supporting one leg of the callipers on the finger as seen in the diagram *Fig. 4*. The micrometer is then adjusted until the opposite leg of the callipers is in light contact with the micrometer spindle face.

The 'feel' is best obtained by rocking the free leg of the callipers past the micrometer spindle in the direction of the arrows, using the anvil of the micrometer as the fulcrum for the movement.

The Slide Gauge

The serious worker will have provided himself with at least one micrometer and possibly two, so that he may make measurements up to 2 in. diameter. Above this size the most suitable equipment is the slide gauge. At one time these instruments were very expensive; today, however, excellent tools at a reasonable price are available. When it is considered that a 6 in. slide gauge can take the place of several micrometers, and can be used for both outside as well as inside measurements, the actual cost is probably well worthwhile.

Fig. 8. The micrometer depth gauge

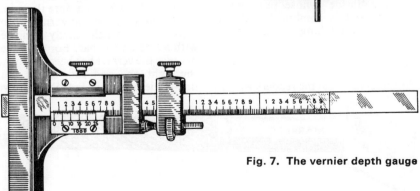

Fig. 7. The vernier depth gauge

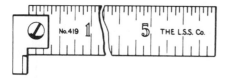

Fig. 9. The hook rule

A typical slide gauge is illustrated in *Fig. 5*.

The slide gauge consists of a steel rule having a fixed jaw at one end, and a sliding head with a second jaw free to move up and down the rule. When these jaws are closed on a piece of work a datum line and vernier indicate the dimension on the rule scale.

The fixed jaw is made in one piece with the rule, while the sliding head itself combines several parts, the jaw being a portion of the head having the vernier scale attached to it. As it is not possible to set the tool with any accuracy simply by sliding the head along by hand a fine adjustment is provided. This consists of a clamp, knurled nut and screwed stud the latter being rigidly attached to the moving jaw. By locking the clamp and turning the nut the jaw can be slid in or out of engagement with the work until the correct setting is found. The jaw itself is also provided with a lock for use when needed.

Two forms of jaw are in common use: the 'knife edge' for making external measurements in narrow recesses and the 'inside' whose purposes is obvious and is combined with the main jaws for taking outside dimensions.

How to Read a Vernier

As there may be some readers who do not understand how to read a vernier a short explanation has been included here.

The bar of the tool to which the vernier is fitted is graduated in fortieths-of-an-inch or 0·025 in., every fourth division representing a tenth-of-an-inch being numbered. The vernier plate, attached in this instance to the sliding jaw of the gauge is divided into 25 parts and numbered 0·5, 0·10, etc., up to 25. These 25 parts occupying the same space as 24 divisions on the bar.

The difference in width between one of the 25 spaces on the vernier and one of the 24 divisions on the bar is, therefore, 1/25th of 1/40 or 1/1,000 of an inch. If the tool is set so that the zero (0) line of the vernier and the zero (0) line of the bar coincide, the next line to the right on the vernier will differ from the zero line on the bar by 1/1,000 of an inch, the second line by 2/1,000 of an inch and so on, the difference continuing to increase by steps of 1/1,000 of an inch until the line 25 on the vernier coincides with line 24 on the bar.

To read the slide gauge note how many inches, tenths (or 0·100 in.) and fortieths (or 0·025 in.) the zero mark on the vernier is away from the zero mark on the bar; then note the number of divisions on the vernier from zero to a line which exactly coincides with a line on the bar. For example: in *Fig. 6* the vernier has been moved to the right one and four-tenths plus one-fortieth inches (1·425) as shown

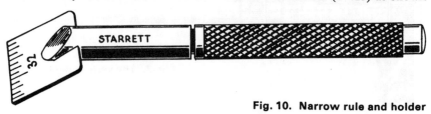

Fig. 10. Narrow rule and holder

MEASURING EQUIPMENT

on the bar and the eleventh line on the vernier coincides with a line on the bar. Eleven-thousandths of an inch must, therefore, be added to the previous reading of 1·425 in. making a total of 1·436 in.

The same procedure applies to height or depth gauges incorporating a vernier.

Depth Gauges

The depth gauge illustrated in *Fig. 7* makes use of the vernier principle. It has one advantage over the built-in depth gauge found in some vernier callipers, in that its base is wide and can be held down firmly during the measuring operation, moreover, the width of the base makes certain that the rule itself is held vertically. It is a tool much in use industrially, because it may be used rather more speedily than the micrometer gauge illustrated in *Fig. 8*. This gauge is also provided with a wide base and has interchangeable depth rods that allow a wide range of measurements to be taken.

On the score of expense, however, the amateur worker will be well served by a slide gauge having a depth gauge incorporated. But he must be prepared to take some trouble in making his measurements.

Steel Rules

We shall be hearing more of the steel rule in the succeeding chapter on marking off. However, no description of measuring equipment could be considered complete without a mention of two particular types of rule having somewhat special applications.

The first of these is the hook rule illustrated in *Fig. 9*. It enables the user to take the guess work out of setting a rule perfectly level with the end of a piece of work and is also useful when adjusting either the dividers or the inside callipers to some definite measurement. Rules of this type are often narrow and proportionally thick. They are sometimes hardened, an advantage in maintaining the quality of the engraving.

The narrow rule seen in the illustration *Fig. 10* has a particular interest for the turner or the user of any of the other machine tools to be found in the workshop.

Those who have had only a short experience of the lathe will quickly have realised how difficult it is to make measurements in the somewhat confined space that is usually available. The rules we are considering are, for the most part, a set of four, varying from $\frac{1}{4}$ in. to 1 in. long. They are mounted on a holder and are gripped by a collet contracted by a draw rod operated by the finger nut seen at the end of the handle.

The device illustrated is by one of the more famous of the American fine tool manufacturers and its availability today may be doubtful. However, one can make these pieces of equipment quite easily, cutting up an old rule to supply the scales needed.

The Taper Gauge

Whilst not exactly meriting the description of a rule the taper gauge illustrated on page 50 is used to measure the diameter of holes. These gauges are graduated in 0·001 in. and the example shown has a range from 0·1 in. to 0·5 in. They are very simple to use, all that is needed is to push the proper blade into the hole when its diameter may be read off directly.

CHAPTER 23

Marking Out

MARKING-OUT is the operation of indicating on components their finished dimensions, as well as the location of any drilled or bored holes that may be required, in accordance with the dimensions indicated in the machine drawings; in addition, guide and reference lines are inscribed as a further aid to the machinist.

The purpose of this process is twofold; firstly, to ensure that a component such as a casting will allow for finishing to the required dimensions, and will not have to be discarded when partly machined; and secondly, the dimension lines, by indicating the prescribed limits, will hasten the machining of the part, and will also guard against the perpetration of errors and spoilt work.

Marking-out prior to machining is the normal and recognised procedure in the small machine-shop, where articles are made singly or in small numbers, but this operation, however well performed, cannot attain absolute accuracy any more than can the subsequent machining processes. The final degree of accuracy of the finished part will be the outcome of the inaccuracy due to marking-out added to that of machining, and any cancelling-out of these errors will depend only on chance.

In production work, marking-out is dispensed with and instead, the machine carries out the work with great accuracy on the component located by means of a jig or fixture: and in this way even the inherent inaccuracy of the machine may be eliminated by the use of guides and guide bushes.

Even in the small workshop, marking-out is not employed prior to the machining of components such as gear wheels, for here the teeth are machine indexed, and their form is determined by a machining operation, which ensures greater accuracy than could be achieved by working to dimension lines.

The Datum Line or Surface

As in the case of a mechanical drawing, a base line is used in marking-out when scribing dimension lines and locating centre points on the work.

When marking-out sheet metal, the process is very similar to that used on the drawing board, but, in the case of castings and objects of more solid form, a surface and not a line is usually necessary for reference.

In the first place, therefore, a flat surface should be prepared by filing or machining, to afford a stable base on which the component can stand on the surface plate or other plane surface.

From this datum or reference surface all dimension lines and location points should be marked-off by means of a surface gauge used where necessary in conjunction with other appliances on the surface plate or marking-out table.

When marking-out the centres of a series of holes, for example, it is important that each in turn should be located from a single datum point; for if, on the other hand, each centre is marked-off from the previous centre and there is an error of, say, five-thousandths of an inch in the setting of the dividers, there will then be a cumulation of this error, and the

MARKING OUT

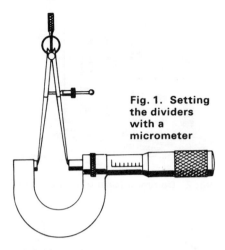

Fig. 1. Setting the dividers with a micrometer

tenth hole of the series will be located forty-five-thousandths out of place, whereas, when each hole is located from a common centre the error should never exceed five-thousandths of an inch.

Stages in Marking-out

In many instances it would be impracticable to transfer all the details of the machine drawing to the work at one setting and the usual practice is, therefore, to provide such marking-off as is necessary for the machining immediately required.

For example, when machining the casting of an engine cylinder, the first operation is usually to bore and face the part. In this case, all the marking-out required is the indication of the centres from which the machinist can set up the work, and, in addition to the true dimension line, a witness line should be scribed to indicate the limit of the preliminary boring operation.

After the casting has been bored and faced, it is again marked-out on the machined surfaces to show the position of the stud and bolt holes.

When preparing plate or sheet material for machining, both the dimension lines and the drilling centres should be marked-out in the first instance, for not only is this generally more convenient, but the preliminary machining operations by obliterating the datum line may render further marking-out difficult to accomplish.

Errors in Marking-out

If the datum surface is not truly flat, the component may rock on the surface plate and thus assume two different positions; this is liable to cause inaccuracy and confusion, as the marking-out may, in this case, be located from two instead of from a single datum surface.

In the same way, a burr on the datum surface arising from careless handling, or a metal chip, may cause rocking or displacement of the work with attendant inaccuracy during marking-out.

Errors may also arise from the use of defective tools, as when the joints of callipers and dividers or the adjustable parts of the surface gauge are in need of attention, and are liable to move when in use.

Mishandling of these tools, and failure to keep them in a place of safety on the bench, may also result in their adjustment being upset unwittingly.

A convenient and accurate method of adjusting the dividers, and one often used by mechanics, is to make the setting by means of a micrometer as shown in *Fig. 1*.

In order that only the points of the dividers may make contact with the micrometer's measuring surfaces, the tips should be finely tapered, or, better still, the outer surface of the points should be ground flat, and only the inner and side surfaces are stoned when the points are sharpened.

Even if, in the first instance, the setting of the dividers is found to be incorrect, by this method very fine adjustments can be made when a process of trial and error is adopted.

THE importance of using a datum line or surface in marking-out has already been stressed, and in the case of sheet material this line may suffice as a base for all subsequent marking-out operations comprising two-dimensional lining-out, but when dealing with work of more solid form, a datum surface is necessary to enable the marking-out to be undertaken in three planes in space, thus representing three-dimensional delineation.

Usually this datum surface or base will not of itself afford adequate guidance for the marking tools employed, and it becomes necessary to extend this reference base by resting the work on a large flat surface, from which all points and dimensions can be located.

Fig. 2. Common form of angle plate

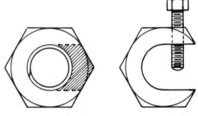

REMOVE THREAD FROM NUT & CUT AWAY SHADED PORTION

DRILL & TAP FOR SCREW AS SHOWN

Fig. 3. Clamp made from a nut

Marking-out Tables

The extent of the surface employed must be adequate to accommodate the work, together with the appliances used for marking-out. In the case of large castings, a marking-out table with planed cast-iron top and edges is used, but for smaller work a sheet of plate-glass or a surface plate may be employed.

For this purpose the machine-finished surface plate is quite satisfactory, but in the small workshop the hand-scraped surface plate is often employed in this way, in addition to its normal use as a reference surface.

Although the surface plate is usually of massive construction and is heavily ribbed to afford rigidity, it is important that it should be protected from damage when not in use, and on no account should it be employed as a hammering block, for any burrs or distortion so caused will inevitably lead to inaccuracy when marking-out is undertaken.

When large work is dealt with, the table should be accurately levelled so that a spirit-level can be employed to facilitate setting and marking-out the component.

Angle Plates

Usually a component can be marked-out while resting on its datum surface on the surface plate but, on the other hand, at this stage there may be no machined datum surface, and the part is then attached to a fixture such as an angle plate, which itself furnishes the datum surface. An example of the type of angle plate in common use is depicted in *Fig. 2*.

These fixtures are provided with holes or slots for attaching the work by means of bolts or dogs, but at times clamps are used for this purpose; these clamps may be of the ordinary G variety or of the toolmaker's pattern, but for securing sheet material the type shown in *Fig. 3* will be found

MARKING OUT

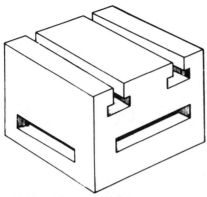

Fig. 4. Box angle plate

convenient, and the method of making them will be readily gathered from the drawing.

The box angle plate, shown in *Fig. 4* with its six machined surfaces, is a more elaborate and more generally useful form of fixture, for, with the work attached to one face, the five remaining surfaces may be used at will for positioning the work, and it follows that at a single setting a component can be marked-out in three planes at right angles. In some patterns of these angle plates T-slots are provided to facilitate the work of clamping.

The adjustable angle plates and tilting tables, largely used in connection with drilling and milling operations, may be employed for setting work on the surface plate when marking-out.

Packing Strips and Raising Blocks

Some difficulty may at times be experienced when mounting irregular work on the surface plate, for example the machined datum face may carry a tenon as in the case of a tailstock sole-plate, and two similar packing pieces will then be required to support the work.

To ensure equality of thickness, these parallel packing pieces are ground on all their longitudinal faces, and in addition the surface may be chromium plated to afford protection from damage and rusting.

Where critical adjustment of the height of setting is required, adjustable packing strips, or adjustable parallels as they are called, can be used with advantage for these devices can be accurately set to any required height by means of a micrometer. From the drawing in *Fig. 5* it will be seen that these parallels are composed of two grooved members, sliding on an inclined plane and capable of being locked in any position by a setscrew.

Screw Jacks

Small screw jacks and at times wedges, are largely used for setting work on the marking-out table, either in the level position or inclined at an angle to the surface. *Fig. 6* shows one of a pair of screw jacks that were specially made for adjusting the position of heavy castings on a large marking-out table.

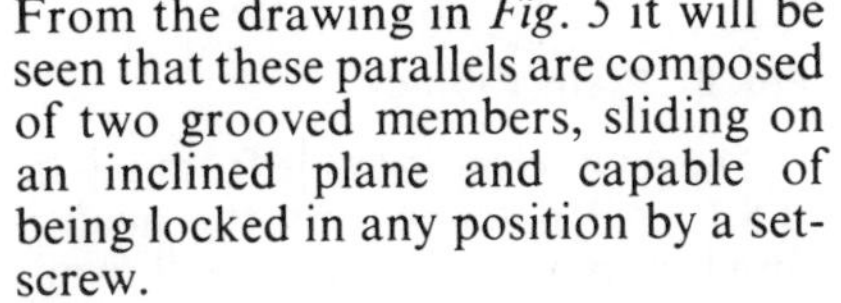

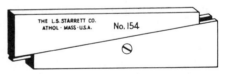

Fig. 5. Adjustable parallels

Fig. 6. Screw Jack

Fig. 7. Precision V-block

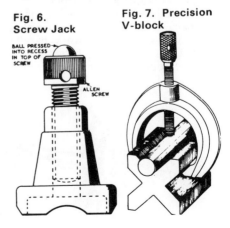

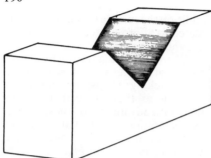

Fig. 8. Plain V-block

The jack screw is made from a ½-in. B.S.F. Allen screw, and the head is fitted with a ⅜ in. bearing ball firmly pressed into place; but for smaller work a screw jack of half this size with a ¼ in. screw and a 3/16 in. ball will be found more convenient.

To reduce working friction, particularly when heavy castings are dealt with, the base member of the jack may be fitted with a bronze bush threaded to engage the jack screw.

The head of the jack should be cross-drilled for the insertion of a tommy bar, to give increased leverage and to facilitate adjustment in awkward positions.

V-Blocks

V-blocks of cast-iron or hardened steel are used to support and locate round work on the surface plate.

These blocks should be purchased in pairs, to ensure that they are alike and will support the work parallel with the surface on which they rest.

The four V-grooves, with which

Fig. 11. The scribing gauge

blocks of the type illustrated in *Fig. 7* are provided, afford a ready means of setting the work at various heights, and the clamps may be used to secure the work in position during marking-out.

The larger V-blocks, which are also machined in pairs, usually have only a single open V for supporting round material of large diameter. A large plain V-block is illustrated in *Fig. 8*.

Surface Gauges or Scribing Blocks

These devices are used for scribing lines on plane or irregular surfaces at a predetermined distance from the base surface, and in addition they may be employed to set components in parallel alignment with the surface of the marking-out table.

In the universal type of gauge illustrated in *Fig. 9*, and recommended for general use, the heavy base member has a truly flat under-surface in

Fig. 9. The surface gauge

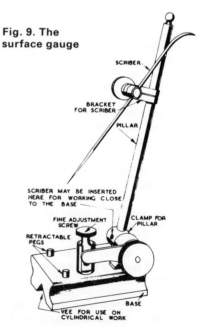

MARKING OUT

which a V-groove is machined to afford a location on cylindrical work.

In addition, the base is furnished with retractable guide pegs, by means of which the appliance can, if desired, be guided from the edge of the surface plate or other machined surface. The smaller forms of surface gauges are usually equipped with a removable guide piece for clamping to the base, and the guide pegs are then omitted.

The two-ended scriber is carried in a bracket which is clamped to the vertical pillar of the gauge.

This pillar is in turn clamped in a cross-drilled spindle, which can be rotated to impart a radial movement to the pillar for adjusting the height of the scriber point.

Fine adjustment of height is effected by means of a rocking gear, comprising a radial arm attached to the pillar clamp and controlled by a fine thread finger-screw.

If desired, the pillar can be removed and the scriber mounted directly in the pillar clamp when working close to the base of the gauge. As in the previous case, fine adjustment of the position of the scriber point is effected by means of the screw-controlled rocking gear.

As previously mentioned, the surface gauge may be used for levelling and positioning work on the surface plate; for this purpose the curved end of the scriber is turned downwards, and is then applied to various points on the work under adjustment until uniform contact is obtained.

In addition to its use as a setting gauge, the surface gauge is more often employed for scribing lines parallel with a reference surface or edge.

It is essential that an accurate method of setting the scriber point should be adopted, when the surface gauge is used to scribe lines or locate points at a definite distance from the surface of the marking-out table. Whenever a rule is used for this purpose, an angle plate or a special rule-holder should be employed, to ensure that the rule is maintained in a truly vertical position while the setting is made.

The device shown in *Figs. 10* and *10A* which was specially designed for this purpose, will hold securely and accurately any rule of standard width without obscuring the graduations.

The accurate setting of the scriber point against the rule will be facilitated if the scriber itself is positioned horizontally, and a magnifying glass is used when making the final adjustment.

Distinct but lightly cut lines should be made when scribing with the surface gauge, for any attempt to cut deeply into the metal will result in inaccurate marking due to springing or displacement of the scriber, although the rigidity of the scriber will be enhanced if it is clamped in its

Fig. 10. The rule stand

bracket with the minimum of overhang.

The Scribing Gauge

This device, which is depicted in *Figs. 11* and *11A*, will be familiar to woodworkers as the scratch gauge, but in this instance some modifications have been introduced to make it suitable for metal work.

Its compactness renders it more convenient than the surface gauge for scribing lines parallel with a reference edge, and the length of its base, unlike the jenny callipers, ensures that true parallelism at an exact distance from the base line is maintained, particularly in the case of sheet metal work.

As will be seen in the drawing, a sliding scribing bar, clamped in the base member, carries a removable scriber point, which is adjusted to set the bar in a horizontal position when the fence is in contact with the work.

To avoid its digging into the work, the scriber point should be set with a slight inclination to give a trailing action.

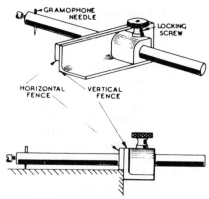

Fig. 11A. The scribing gauge

Fig. 10. The rule stand

Fig. 10A. The rule stand

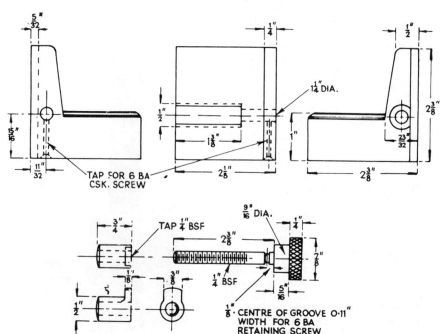

Squares

Squares are used in marking-out either when applied to an edge as in sheet metal work, or standing on the surface plate for marking vertical lines, or to set surfaces and scribed lines in the vertical position.

Needless to say, only accurate squares of good quality should be employed for this purpose, and to minimise wear the blade should be hardened.

The edge of the blade should be plain, and not double-bevelled as in some types of toolmaker's squares, otherwise errors of marking-out may arise owing to the position of the line scribed varying with the angle at which the scriber is held.

For convenience, and to avoid encumbering the surface of the marking-out table, squares of various sizes should be provided for use with large and small work.

The Combination Square

Although the combination square can be used as an ordinary try-square, it has the further advantage that the stock or head slides on the grooved blade and can be clamped to it at any desired point. Combined with the stock of the square is a spirit-level and a 45-degree mitre-square. If the marking-out table is set truly level,

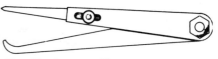

Fig. 13. Jenny callipers

this spirit-level may be used to set components parallel with the surface of the table.

Since the blade can be clamped at any point in the head, the combination square makes a useful parallel marking gauge, and in addition it may be used as a depth gauge when marking-out. For the sake of convenience, a scriber is held frictionally in the head and ready for immediate use. When the centre head, shown in *Fig. 12* is substituted for the ordinary head, the tool can be used to find the centre of a shaft or other cylindrical part.

The addition of a protractor head completes the usual equipment of the combination square, and enables angular work to be set out on the marking-out table.

Dividers

When this tool is used for spacing the distance between points or for scribing circles, one leg should be located by means of a small punch dot in order to avoid errors due to change of position during the operation.

Jenny, Hermaphrodite or Odd-legs Callipers

This tool is generally used for scribing lines parallel with and at a predetermined distance from a reference edge on the work.

As a rule, these callipers are made with a plain frictional joint, as shown in *Fig. 13* but some mechanics prefer the type fitted with a screw adjustment and illustrated in *Fig. 14*. As the latter form cannot be purchased, they have to be made from a pair of screw-adjustable inside callipers. One leg is

Fig. 12. The centre head

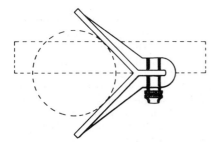

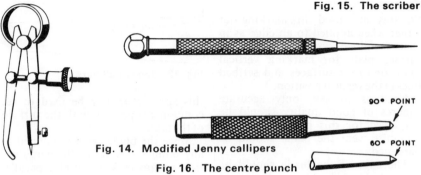

Fig. 15. The scriber

Fig. 14. Modified Jenny callipers

Fig. 16. The centre punch

shortened and a piece of steel to carry the set-screw and hold the scriber point is brazed on; the other leg is then heated and turned inwards as shown in the drawing. The scriber point is best made from a new gramophone needle.

When adjusting the plain type, the point is engaged with the required graduation on the rule and the other leg is gently pressed into contact with the end of the rule.

If the joint of the callipers is in order this setting should be exactly maintained when the pressure of the fingers is released, but, if the contact leg tends to spring away from the end of the rule, the joint requires attention. The screw-adjustable type is easily set by placing the contact leg against the end of the rule and then bringing the scriber point into register with the graduation required.

Scribers

Although it might be thought that neither the design nor the use of the scriber requires description, this is far from being the case; for a faulty scriber or its improper use may cause serious errors in marking-out.

The fine sharp point of the scriber is easily damaged, especially when carried in the tool kit or allowed to fall from the bench; for this reason the toolmaker's scriber, shown in *Fig. 15* is provided with a reversible point which fits within the handle and is retained in place by a screw chuck.

As a further safeguard, the upper end of the handle is made of hexagon form to prevent the tool from rolling on the bench.

As in the case of dividers, the scriber should be sharpened periodically, for the pressure which it is necessary to apply to a blunt scriber may cause displacement of the guide rule with consequent inaccuracy during marking-out operations.

Centre Punches

For the initial marking of centre points, a centre punch with a fine point of some 60 degrees included angle should be used, and in the case of drill holes, this should be followed by a punch of 90 degrees or more, *Fig. 16*, to afford adequate location for the drill.

It is important that the marking punch should have a really sharp point, otherwise it will be found that no proper centre is formed to locate the point of the dividers when spacing out dimensions or scribing circles.

When marking-out, centre points are usually located by the intersection of two scribed lines, and it is here essential that the punch point should be exactly placed; this will be greatly facilitated by employing a magnifying

MARKING OUT

glass, and the punch must be held truly vertical to ensure that the centre is not drawn over to one side when the punch is struck with the hammer.

The hammer used for this purpose should be of light weight, but well balanced, to enable decisive but well-controlled blows to be struck.

As great accuracy is demanded of the fine tools used for marking-out, their careful preservation will be well repaid. The small kit of hand tools required may be kept in a separate box or drawer near the marking-out table and ready for immediate use, whilst the larger tools and appliances should be stored together in a cabinet so that they are readily accessible, but are at the same time well protected from possible damage.

Preliminary Mechanical Work in Marking-out

When components are made singly or in small numbers, some preliminary work is usually required to prepare them for the actual marking-out process. In the case of castings of malleable iron, brass and aluminium alloys, as well as forgings, some setting or straightening may be required to give the parts their correct form, otherwise the excessive removal of metal during machining, in an attempt to remedy the distortion, may well render the casting useless.

In addition, as the component is not positioned for machining in a jig, it is usually necessary at the outset to provide a flat reference or datum surface for locating the part on the marking-out table, and later to position it for machining.

In the case of sheet metal work, however, a straight base edge may be formed, from which centres can be located and dimensions marked-off as in making a machine drawing; this edge can also be used later to locate the work for machining operations.

In the above example the base line serves for two-dimensional marking-out, but where solid objects are concerned three-dimensional marking may be required and a datum surface becomes necessary in most cases.

The datum surface is prepared either by machining or by filing and scraping, and it is essential that it should be truly flat to avoid any possibility of rocking on the marking-out table; at the same time it must be so positioned as to ensure that the dimensions specified in the machine drawing can be accommodated on the component when marking-out. Sometimes the datum surface is not used directly, but the component is located by this surface on an angle plate or other fixture during both the marking-out and machining operations.

In this manner, as has already been mentioned, when the part is attached to a box angle plate, five of its surfaces can be marked-out or machined in accurate rectangular relationship without alterations of the setting.

In addition to sheet metal work and solid objects, a third type of component, represented by a hollow cylinder, frequently has to be marked-out. For example, the casting of an engine cylinder with cast-in bore must be marked-out to enable the bore and flanges to be machined, prior to marking-out and machining the stud holes in the cylinder flanges.

After it has been ascertained that the bore is centrally placed in the casting, the centre line of the bore is marked-out as a guide to the machinist when adjusting the casting for the boring operation. To enable this to be carried out, plugs of metal or hard wood are fitted to the bore at either end, and the bore centres are marked on these by means of dividers or jenny callipers; from these centres dimensional and witness circles are scribed as a guide to the operator.

In the case of a large cylinder,

bridge pieces as shown in *Fig. 17* are usually fitted.

Marking-out Media

After the casting has been pickled and cleaned, and the datum surface has been truly formed, the surface of the component is prepared so that the lines formed by the surface gauge or scriber will not only be clearly visible, but will remain so for the guidance of the mechanic during the subsequent machining operations.

Immediately before treating the surface of the work for marking-out it should be thoroughly cleaned by the use of fine emery cloth, or by wiping over with petrol or lighter fluid, to remove oil and finger marks.

Large rough castings are usually painted with a light distemper, or with a mixture of glue and whitening; although it will be found that a mixture of whitening, white lead, or zinc oxide and French polish or spirit varnish will not only be quicker drying, but will adhere more firmly to the metal.

For preparing any metal for marking-out, lacquers are now obtainable, which are usually deep blue in colour, and have the advantages of being extremely quick-drying as well as adhering tenaciously to the surface of the work.

Although impervious to oil, these lacquers can be readily removed by the application of a rag dipped in methylated spirit.

For the sake of convenience, a small quantity of the lacquer should be kept in a bottle fitted with a cork carrying a brush and should be applied to the work in a thin even coat.

Use of Jenny Callipers

The jenny calliper and its usual methods of adjustment have been described in some detail earlier, for this simple tool is perhaps the

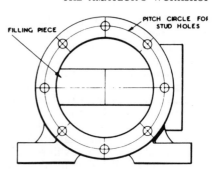

Fig. 17. Using bridge pieces

most generally used appliance for the quick and ready marking-out of simple dimensions, and for locating drilling centres in much of the work encountered in the small machine-shop. Moreover, this tool can also be used for marking lines parallel with a datum edge, finding the centre of a shaft, marking the pitch circle for the stud holes in a cylinder cover, and other work of a similar nature.

Scribing Lines Parallel to a Datum Edge

The simplest operation for which the jenny is used is the scribing of a line parallel to the edge of a piece of plate material. Such a line may be required for the location of drilling centres, or it may be used as a witness line when filing. The method employed to scribe a parallel dimension line is illustrated in *Fig. 18* and, as one leg of the jenny has a guide or contact surface set at a predetermined distance from the

Figs. 18 and 19. Using Jenny callipers

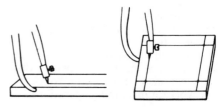

MARKING OUT

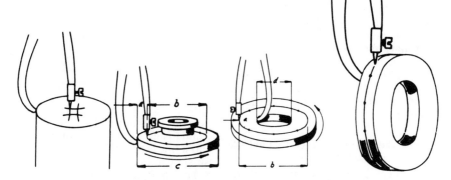

Figs. 20 and 21. Using Jenny callipers

Figs. 22 and 23. Using Jenny callipers

scribing point, it is essential that a true face or edge should be provided to ensure the accurate transfer of dimension to the work.

Marking-out Centres for Bolt Holes

The operation of marking-off the centres of four bolt holes on a square piece of material is illustrated in *Fig. 19*. In this case, the jenny is set to the distance of the centre of the holes from the edge of the component, and the intersections of the four scribed lines indicate the drilling centres for the bolt holes. It will be appreciated that this principle may also be used to mark-off additional holes on the same component, if the jenny is reset to the distance of the new centres from the edge of the work.

Finding the Centre of a Shaft

One of the more common uses of the jenny callipers is to determine the centre of a round shaft. To do this, the jenny is set to the approximate half-diameter of the shaft and, as shown in *Fig. 20* four arcs roughly at right angles to one another are scribed on the end of the shaft.

The centre of the small area enclosed by these arcs is the centre of the shaft, and this can be accurately determined by resetting the jenny until it is found that the arcs meet exactly at a central point.

Pitch Circles

It is sometimes required to scribe the pitch circle of a series of holes on a component, such as a cylinder cover, which has a circular periphery or register. This is readily done by using the jenny callipers, which are set equal to the distance between the pitch circle and the periphery of the cover.

Fig. 21 illustrates the method of setting and using the callipers, and it will be seen that the dimension $A = \dfrac{C-B}{2}$

Marking-out will be facilitated in this instance if the work is rotated, as shown by the arrow, and the jenny is held stationary; otherwise, the scribing point will tend to rotate about the guide leg and a circle of varying diameter may result.

It may, in some cases, be found more convenient to scribe from the inner machined surface or bore of a component, and the method here used is similar to the foregoing, except that the jenny is set equal to the distance between the pitch circle and the bore, as shown in *Fig. 22*, where $A = \dfrac{B-D}{2}$

In the operation illustrated in *Fig. 23* the jenny is used to scribe a pitch circle on the edge of a disc or shaft, and the method is like that employed for marking-out a line parallel to the edge of a component, as shown in *Fig. 24* but in this case it is more readily carried out if the work is rotated while held in the chuck on the lathe mandrel.

Using the Surface Gauge to Find the Centres of a Shaft

In this case, as is shown in *Fig. 24*, the shaft is supported in a like pair of V-blocks on the surface plate, and the latter then provides the extended datum surface for the guidance of the surface gauge or other appliances.

The V-blocks should be positioned as far apart as possible in order to reduce any error of alignment arising from lack of straightness of the shaft, and it may be found that clamping the shaft in one V-block will facilitate the marking-out operation.

As illustrated in *Fig. 25* the height of the upper surface of the shaft is set on the surface gauge and then measured by a rule supported in a stand; if necessary the height of the rule should be adjusted to indicate an exact fraction of an inch at this point.

Next, the diameter of the shaft is measured either directly by means of a micrometer or callipers, or the height of the lower surface of the shaft, obtained by using the surface gauge and rule, can be subtracted from the height of the upper surface to deter-

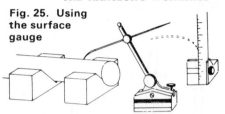

Fig. 25. Using the surface gauge

mine the shaft diameter. The scriber point is then lowered by an amount equal to half the diameter of the shaft, and a line is scribed across the end of the shaft as in *Fig. 26*.

The shaft is now rotated in the V-blocks through an angle of approximately 180 degrees, and the scriber point of the surface gauge is brought into contact with one end of the scribed line. The scriber is then applied to the other end of the line without moving the shaft, and, if the coincidence is exact, the scribed line passes through the centre of the shaft, but if not, the scriber must be adjusted to halve the discrepancy and a new line is traced. This new setting should be checked by again rotating the shaft and applying the scriber to either end of the line.

After the centre line has been truly scribed, the shaft is rotated through an angle of 90 degrees by using a try square in conjunction with the scribed centre line as shown in *Fig. 27*, and a second centre line is then scribed with the surface gauge at right angles to the first.

The intersection of these two centre lines indicates the centre of the shaft.

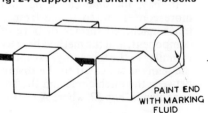

Fig. 24 Supporting a shaft in V-blocks
PAINT END WITH MARKING FLUID

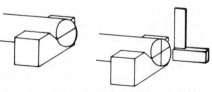

Fig. 26. Using the centre line

Fig. 27. Using the square

Marking-out for Squaring the End of a Shaft

The method used includes the geometrical construction described for marking a square on the end of a shaft.

Paint the end portion of the shaft with marking fluid, and with the jenny callipers scribe a line 'A', to denote the length of the square, by rotating the shaft in V-blocks on the surface plate. The shaft is then clamped in one of the V-blocks to prevent further rotation. Find the centre of the shaft as already described, and with a mitre square or protractor scribe a line at 45 degrees through the centre. Set the scriber point of the surface gauge to one end of this line, and scribe lines on either side of the shaft up to the line 'A' and also across the end of the shaft. Set the scriber to the other end of the 45 degrees line and scribe three further lines in the same manner. The four lines thus scribed on the sides of the shaft will denote the corners of the square required, and the figure is completed on the end of the shaft by using a try square.

If a square with a diagonal smaller than the shaft diameter is to be formed the shaft should be turned to the correct diameter prior to marking-out. The length of the diagonal of a square, and hence the shaft diameter required, is 1·41 times the length of its side.

Use of the Surface Gauge

When marking-out a keyway on a shaft by means of the surface gauge, the shaft is clamped in V-blocks on the surface plate, and the horizontal centre line is scribed across the end of the shaft by the method already illustrated in *Figs. 25* and *26*. The scriber point is then set in turn, by means of the rule, to half the width of the keyway both above and below the centre line, and lines are scribed across the end of the shaft and along its sides for a distance equal to the length of the keyway.

Both the length and the depth of the keyway are marked with the jenny callipers, and these dimensions are then marked-out by using the try square standing on the surface plate, *Fig. 27*.

When two keyways at an interval of 180 degrees have to be marked-out on opposite sides of the shaft, the same procedure is adopted, but in this instance the second keyway is marked-out by applying the surface gauge to the opposite side of the shaft at each setting of the scriber point.

Angular Location of Keyways

A number of keyways or splines can be marked-out by rotating the shaft in the V-blocks and setting its centre line to the appropriate angle by means of a protractor *Fig. 28* and *Fig. 29*. By this method, too, a keyway or keyseat can be marked-out in any required angular relationship to the centre line of the shaft, or to another keyway, as is sometimes necessary when cutting keyseats in a petrol engine half-time shaft for locating a cam or a gear pinion.

Marking-out Internal Keyways

Internal keyways in flywheels, gear wheels and collars can also be marked-out in the same way after the machining of the component has been completed.

Smaller components such as gear wheels can usually be fitted to an arbor supported in V-blocks when

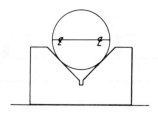

Fig. 28. Setting out for marking keyways at an angle

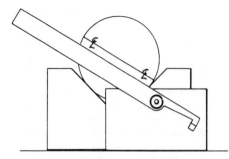

Fig. 29. Setting out for marking keyways at an angle

marking-out internal keyways, and in this case also it may be necessary to maintain a definite angular location from a gear tooth, or an integrally formed cam to provide, for example, for the correct setting of the valve timing in a petrol engine. The relative location of these components will be shown in the machine drawings, and these details should be closely observed when marking-out to ensure the correct assembly of the finished parts.

The methods described for locating keyways apply also to the positioning of cross-drilled holes in shafts and the location of dowels and register pins.

Marking-out a Duplicate Part

When a component is broken or becomes unfit for further service, the question arises of making a copy of the part for use as a replacement. Needless to say, where the original part is of simple design and is but little defective

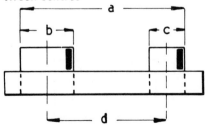

Fig. 30. Determining the distance between centres

it can, in some instances, be used as a template or guide both for machining the overall dimensions and for drilling any holes required.

In general, the first procedure is to select a datum surface or surfaces from which the dimension of the part and the location of bolt and bearing holes can be determined. Following this, either an accurate drawing can be made and the dimensions transferred therefrom to the work, or, in the case of parts of simple constructtion, these dimensions can be markedout directly on the material preparatory to machining.

The actual dimensions of a part are measured either by rule, callipers, depth gauge or micrometer according to the degree of accuracy required, and the dimensions of a duplicate part are likewise checked after marking-out and machining have been completed.

The position of a hole in relation to a datum surface can be conveniently determined by fitting a short piece of turned rod to the hole, and then measuring the distance of the circumference from the datum surface with either a plain or a micrometer depth gauge, and subtracting therefrom half the diameter of the rod.

To determine the distance between the centres of two holes, pieces of rod should be fitted as in the previous example, and, as illustrated in *Fig. 30* the overall dimension 'A' is measured and half the length of the diameters 'B' and 'C' is then subtracted to give the inter-centre distance 'D'.

Reference to *Fig. 31* will show a method of setting-out hole centres, whose distance apart has already been measured.

Determine the distance of both centres from the horizontal datum surface as indicated by the lines 'A' and 'B', and scribe these centre lines by means of the jenny callipers or scribing gauge. Scribe the vertical

MARKING OUT

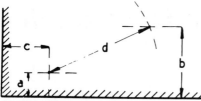

Fig. 31. Marking-off hole centres

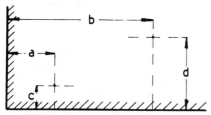

Fig. 32. Alternative method of marking-off hole centres

centre line of the first hole from the vertical datum surface, and from the point of intersection, which indicates the hole centre, scribe an arc with a radius 'D' equal to the intercentre distance of the two holes; this second point of intersection will indicate the centre of the other hole.

An alternative method of marking-out these hole centres is illustrated in *Fig. 32*. The vertical centre lines or co-ordinates are first drawn at the correct distance apart and the horizontal centre lines are then scribed to locate the hole centres. The distances 'A' and 'B' of the hole centres from the vertical datum surface are measured with a depth gauge as already described, and if necessary the rod is removed from the first hole when the position of the second is being determined.

The co-ordinates are then marked-out in accordance with these measurements. In a similar manner the horizontal centre distances are measured and the points of intersection of the horizontal lines 'C' and 'D' with the co-ordinates indicate the hole centres required.

CHAPTER 24   # The Dial Test Indicator

THE Dial Test Indicator is probably one of the most important items in the tool kit of the serious worker. It enables him to set his work correctly in the machine tool he is using, and also allows him to check the accuracy of the machines and their equipment.

Dial indicators are made in a variety of forms, but those most commonly employed have clock-type faces (hence the colloquial name for the device the 'clock') and are operated by a plunger driving a train of gears set in the body of the instrument. The plunger may project from either side of the indicator as seen in *Fig. 1* or may enter the body from the back as will appear in later illustrations.

The range and capacity of the indicator varies but those of general utility are graduated in one-thousandths of an inch and have a maximum measuring capacity of $\frac{1}{4}$ in.

The indicator is designed to be mounted on a suitable stand such as the 'Eclipse' magnetic base already illustrated in *Fig. 1* or it may be attached, as we shall see presently, to a conventional surface gauge enabling it to be set by means of the rocking lever adjustment commonly provided.

The bezel of the instrument carrying its dial can be rotated, and in some cases subsequently locked, when setting the indicator needle to zero before taking a reading.

The plunger fitted is normally supplied with a single contact, either in the form of a ball in a holder or a larger spherical foot screwed to the plunger extremity. But there are numbers of applications where either of the two feet mentioned would be of little use. The user must then furnish himself with additional feet such as those depicted in the illustration *Fig. 2*.

All three feet are designed to fit over the plunger of the indicator and are held in place by the thumb screw seen in the illustration. The foot depicted on the left is an extension enabling the indicator to reach what might otherwise be an inaccessible surface. Extensions of this nature are sometimes provided with a long pointed stem that may be used, for example, on a narrow projection such as a spigot machined on a component held in the lathe chuck.

Fig. 1. A dial indicator mounted on an 'Eclipse' magnetic base

THE DIAL TEST INDICATOR

The extension in the middle, known as the 'elephants foot', is particularly useful when applied to tapers as will be seen later. The third extension is another elephant's foot able to reach further than the first.

When the indicator is mounted on the surface gauge, as illustrated in *Fig. 3*, it may be used for a variety of purposes such as setting work true in the 4-jaw independent chuck.

Fig. 2. Alternative feet for the indicator

In this instance the pins in the base of the mounting are pressed down and engaged with the ways of the lathe bed to prevent the base from moving.

It is not always convenient to use a large and somewhat bulky mounting for the indicator, particularly when the instrument is employed about the lathe or shaping machine. *Fig. 4* shows three forms of spigot mounting suitable for use under the lathe toolpost. That at 'A' is a fixed spigot designed for gripping in the 4-way toolpost. At 'B' is a telescopic version of the same device that allows the dial indicator to be mounted as close to the toolpost as possible. It consists of a split adapter fitted with a retractable spigot upon which the clamp of the indicator is placed.

Fig. 3. Dial indicator mounted on a surface gauge

The adapter illustrated at 'C' is intended to be slipped under the centre bolt of the toolpost.

It is illustrated again in *Fig. 5* where it is depicted with the dial indicator in place. Also in the same illustration is a small clamp that permits the indicator to be attached directly to a tool mounted in the toolpost. This device is particularly applicable to the small indicators with rear entry plunger. The economy of space and the versatility of application offered by this attachment make it, in our opinion, one of the simplest and best ways of using the dial indicator in connection with machine tools. The details of the clamp are given in *Fig. 6*.

Fig. 4. Three forms of spigot indicator mounting

Fig. 5. The indicator set on the spigot mounting and alternative mount for clamping to a lathe tool

The mounting in *Fig. 7* was designed and made specifically for the

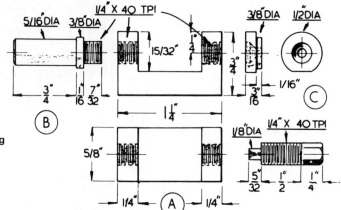

Fig. 6. Details of the mounting for clamping to the lathe tool

Drummond lathe. It fits directly into the feed screw tunnel of the top slide where it is held in place by the expander bolt assembly seen at the right of the device. It will be appreciated that the number of articulations provided make this a very versatile mounting, capable of being attached equally well to the tail of the cross slide or to the back toolpost provided a suitable mounting block is available.

Working drawings of the mounting as applied to the Drummond lathe are shown in *Fig. 8.*

The Internal Attachment

The internal attachment depicted in *Fig. 9*, where the indicator is seen mounted in the 4-way toolpost, can be used for many applications, the first of these being the setting of work truly, either in the 4-jaw independent chuck or on the faceplate, using a reference hole already bored in the part to be machined.

The parts of the attachment, as

Fig. 7. Indicator mounting for the Drummond lathe

THE DIAL TEST INDICATOR

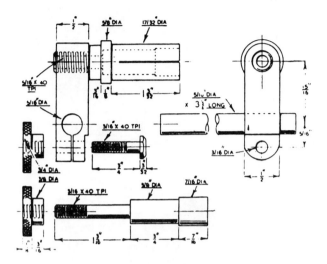

Fig. 8. Details of the Drummond lathe mounting

Fig. 9. The internal attachment for the indicator

Fig. 10. Parts of the internal attachment

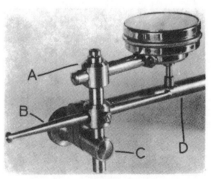

employed with a miniature indicator with back mounted plunger, are illustrated in *Fig. 10*. The 'clock' is gripped in a standard 'A' and is adjusted so that its foot comes directly over the spherical end of the rocking lever 'B'. The assembly is held in the clamp 'C' which serves to secure it to the arm 'D'.

A similar device in use with a side-entry indicator is depicted in *Fig. 11*. In this instance the support 'A' is clamped to the plunger bushing while the standard 'B' holding the rocking lever 'C' can be raised or lowered to adjust the lever in relation to the 'elephants foot' screwed to the end of the plunger.

An indicator with an internal attachment can also be used to set up work on the boring table. Using the indicator in this way entails the provision of a mounting that can be gripped in the chuck as seen in the illustration *Fig. 12*. The ball end of the lever can then be applied to a pilot hole in the work, or to a previously machined edge, turning the internal attachment on its mounting bar in

Fig. 11. Internal attachment for side-entry indicator

order to obtain the most convenient position.

Adjusting the Indicator

The mechanism of the dial test indicator is delicate and no good is done to the instrument by rough treatment such as banging it or lowering it clumsily on to the work. When the indicator is mounted on the surface gauge the user has at his command, as has been said, a fine adjustment that allows the instrument to be brought into contact with the work in a careful manner.

It is not always possible, however, to employ the surface gauge for the purpose so one must make use of any device that can be added to the simple mountings that have already been described. One of these mechanisms is illustrated in *Fig. 13*. This fine adjustment device consists of a body containing a worm and portion of a wormwheel fitted with a spigot to hold the indicator itself. The adjustment is made by operating one or other of the black plastic knobs placed above and below the body of the device.

Fig. 12. Mounting for setting dial indicator internal attachment in the chuck

Fig. 13. Fine adjustment for the indicator

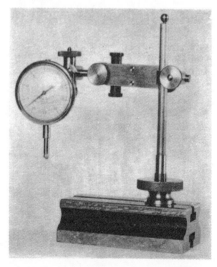

Using the Dial Test Indicator— The 'Wobbler'

An operation that lathe users need to employ perhaps more than any other is the setting of work accurately either in the independent chuck or on the faceplate. For the most part these settings have to be made with reference to a centre marked on the work or else from a hole centre-drilled after the work has been marked off.

Both these reference points entail the use of a piece of equipment called the 'wobbler' in conjunction with a dial test indicator. The wobbler comprises a rod having a well-fitting spring loaded plunger at one end. The plunger is centre drilled at its extremity so

THE DIAL TEST INDICATOR

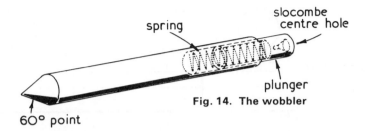

Fig. 14. The wobbler

Fig. 15. Setting the wobbler

Fig. 16. Using the indicator to set a boring bar

that it can rest against a centre set in the tailstock. The rod portion itself has its point turned to an included angle of 60 degrees so that it may be engaged with the centre marked upon the work. The wobbler is depicted in *Fig. 14*. It is used in the manner depicted in *Fig. 15*. With the wobbler in place as previously described the dial indicator is brought into play with its plunger resting against the rod The chuck or faceplate is then turned until the maximum point of eccentricity is ascertained. The indicator is next to set at zero and the lathe again turned until the minimum point of eccentricity is found. This action will also establish where the error lies in relation to the chuck jaws or to a marked position on the faceplate as the case may be. The work may then be moved for a distance of half the indicator reading repeating the process until no movement of the 'clock' needle is observable. The work will then be correctly centred.

The dial indicator can also be used for setting the tool in a boring bar. The illustration *Fig. 16* demonstrates the way in which it is employed. When using the instrument in this way its overall range must be taken into account. For the most part this range is limited to $\frac{1}{4}$ in. Therefore, movements of the tool in the bar must be within this range. For this reason the indicator is probably best employed when setting the cutter for the final two or three boring operations, remembering that increments of

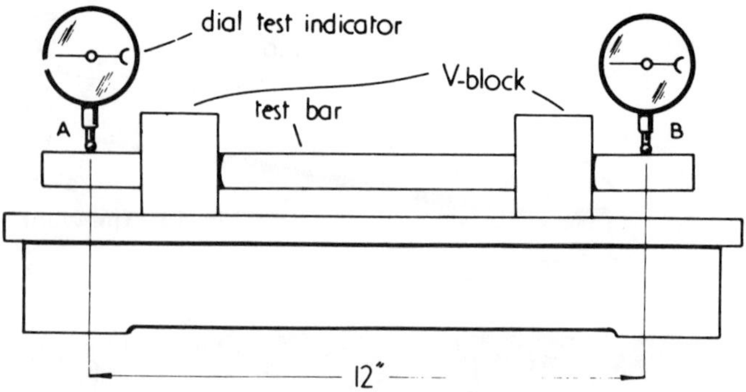

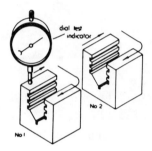

Fig. 17. Checking a pair of common V-blocks

0·001 in. tool movement enlarge the bored hole 0·002 in. each time.

When setting the cutter the indicator must first be zeroed. This is done by bringing the plunger foot into light contact with the point of the tool and then turning the lathe **backwards** until a maximum positive reading is obtained, the bezel can then be turned to bring zero on the dial under the indicator needle. Turn backwards or the point of the tool will score the plunger foot.

The test indicator can be used to check a variety of components such as V-blocks and other equipment. *Fig. 17* shows a method of checking a pair of common V-blocks whilst in diagram *Fig. 18*, a pair of matching blocks are shown being tested to establish their surface accuracy before checking the work seatings.

It is manifestly impossible to list all the possible uses for the indicator. But perhaps two more examples will suffice to encourage readers in the use of the instrument. Both examples concern setting the vertical slide on the lathe. In the first the indicator is applied to the standing jaw of the milling vice *Fig. 19*. In the second example a further use of the internal attachment is depicted. Here, in *Fig. 20*, the attachment is seen applied directly to the work. This method is of value when it is difficult to use the indicator in the normal way.

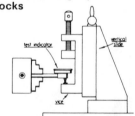

Fig. 18. Checking a pair of matching V-blocks

Fig. 19. Setting the vertical slide

Fig. 20. Setting work mounted in the vertical slide

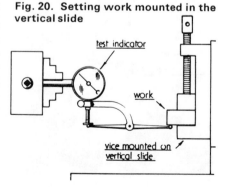

CHAPTER 25

Suds Equipment

THE term 'suds' is applied to the lubricant used when machining various materials in the lathe. At one time this lubricant was a soap-and-water mixture, hence the descriptive name suds. Today, on the other hand, cutting oils of various classes are available; some of these are soluble in water, others are straight oils used without dilution. Industrially, the water-soluble variety has, perhaps, the greater appeal. Our own preference is for a straight oil because the amateur's various machine tools, contrary to those of industry, often have long periods when they are not in use and experience has shown that water-soluble lubricants, despite their makers statements to the contrary, seem to suffer in time from dissociation. The water, settling out, finds its way into slides and this results in discolouration because there is rarely sufficient oiliness to prevent it.

The simplest method of applying lubricant to work is to put it on with a brush. It must be admitted, however, that the process is a somewhat tedious

Fig. 1. Simple method of applying cutting oil

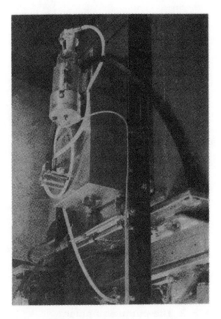

Fig. 2. Suds equipment mounted on the lathe

one. Nevertheless, this can be overcome by making use of a clip to hold the brush against the work and attaching the device to some convenient object such as the back toolpost. By employing Terry clips in the way seen in the illustration *Fig. 1* the brush can be quickly detached for replenishment purposes. A useful device for light cutting but not, of course, commendable when heavy machining has to be undertaken.

There is little doubt that complete suds equipment is the answer to many machining problems. In particular the turning and drilling of stainless steel makes the employment of a copious supply of lubricant essential both for

Fig. 3. The pump equipment

Fig. 5. Mist lubrication system

obtaining a good finish as well as prolonging tool life.

Many years ago we designed and made a set of suds equipment consisting of a reservoir and pump for the lubricant, together with a faucet that could be mounted on the lathe cross-slide, to direct the suds on to the work in hand. This equipment is seen mounted at the back of a Myford lathe, in the illustration *Fig. 2* whilst the pump unit is illustrated in *Fig. 3*. The pump itself is of the rotary vane type directly coupled to a modified motor generator enabling it to be run from the house mains.

The reservoir holds some two gallons of oil and is provided with two wire mesh filters, one on the input side of the pump, the other in the lead from the lathe suds tray. These filters are essential, their efficiency being proved by the fact that the pump, which was, by the way, machined from the solid, has never been dismantled in 15 years and shows no signs of wear.

Fig. 4. The faucet

The faucet *Fig. 4*, is made up from banjo unions as fitted on some automobile carburettors, and has a screw down needle valve to control the flow of lubricant. An ordinary tap would, perhaps, do as well, but the needle valve is somewhat more sensitive.

Oil Mist Lubrication

For those who have a compressed air system installed in their workshops, the device depicted in *Fig. 5* may be of some interest. Originally designed to assist in the machining of large sculptured aircraft components it consists essentially of a spray gun fed with air from the shop compressor and an oil container, holding a gallon or more, that may be located in any convenient place around the machine. The pot seen attached to the gun is in effect a primary reservoir designed to hold a small quantity of oil and promote the depression in the plastic oil pipe leading to the main container.

The gun itself follows the same pattern internally as that designed for paint and described later in chapter 28.

Fig. 6. Miniature mist lubrication equipment

Whilst the system is one of 'total loss', so far as lubricant is concerned, it has been found to be very economical. Moreover it is readily controllable by means of the needle valve fitted to the gun's air-intake so can be adjusted to supply a minimum localised discharge of lubricant.

For those not blessed with an air line the device illustrated in *Figs. 6* and *7* may be of interest. This has its own small compressor and spray gun unit, the latter seen affixed to the cross-slide.

The compressor, a somewhat historic component taken from an early aeroplane, feeds air across the nozzle of the oil discharge pipe in a series of rapid puffs and is driven from the lathe chuck by means of round leather belting, the chuck acting as a pulley.

The compressor itself is mounted on an angle upright forming part of the lathe overhead gear and has provision for belt tensioning.

The oil container holds but a few ounces of oil, but this is sufficient for quite protracted machining, eloquent testimony to the economy of the system.

Fig. 7. Miniature mist lubrication equipment

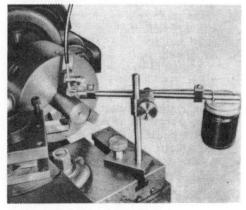

CHAPTER 26   # Lathe Overhead Drives

THE overhead shaft used for driving cutter frames and other tools mounted on the saddle or elsewhere has long been known to lathe users, and its importance to the amateur worker has not lessened with the years.

Formerly, many of the round belt drives, forming the basis of the equipment of the old time lathes, were somewhat elaborate in layout and often cumbersome to set up. These arrangements allowed work to be milled or drilled at either end of the lathe as well as providing a method of machining a long keyway, for example, without the possible necessity of resetting the work.

Today, however, there are other and better ways, perhaps, of achieving the same end, at the same time bringing to bear a good deal more power on the work.

Formerly, in order to provide for extensive lateral movement of the cutter, the driving pulley on the overhead shafting had either to be fitted with a running key permitting the pulley to travel in step with the cutter frame, or the pulley to be a long fixed drum-like affair along which the belt itself could travel as the cutter made progress down the work.

Much of this requirement stemmed from the practice of ornamental turning where long objects such as candle sticks and table legs, needing fluting or otherwise decorating, had to be machined. Today, however, for the most part, the necessity for this kind of facility has gone and it is seldom that the amateur worker needs to move the cutter more than a few inches at a time in a lateral direction, such a movement being easily accommodated by the simple belt layout that can be employed.

Fig. 1 illustrates an elementary overhead drive unit made for use with the Drummond lathe and designed to be bolted to the bench behind the lathe itself. The shafting is carried in a pair of self-lubricating plummer blocks whilst the pulley, adjustable along the shaft, is secured to it by a simple clamping device consisting of an allen grub screw and brass pad bearing on a flat surface machined on

Fig. 1. A simple overhead drive unit for the Drummond lathe

LATHE OVERHEAD DRIVES

Fig. 2. Overhead drive for the Myford ML7 lathe

the shaft itself. A spring loaded tensioning device is mounted on the cross-bar and this fitment, in turn, can be moved and locked where desired. As will be seen, the fitting as a whole forms a convenient support for the lathe lighting unit.

Fig. 2 depicts an overhead drive fitted to a Myford lathe. The frame made from discarded steel bed members, as was that for the Drummond, is bolted to the lathe stand and supports a shaft running, again, in self-lubricating plummer blocks. This shaft is driven from the countershaft supplied with the lathe through a pair of miniature V-ropes.

Two driving pulleys are fitted, one a three-step cone pulley for low speed working, the other a larger wheel used when an increased speed is desired. A simple weight-controlled counterpoise is employed to tension the belt, and the complete set up is seen driving a spotting drill in the illustration *Fig. 3*.

With the advent of the electric motor, and in our case the availability of high-grade low-voltage units at ridiculously low cost, it was possible to combine the drive motor and its miniature countershaft into a unit that could be bolted, either on the back of the saddle or on to the tail of the boring table. In this way, the saddle could travel the full length of the lathe bed, taking with it motor, countershaft and milling attachment with only an electric cable to be accommodated. A unit, of this type, running on 24 volts and giving something of the order of $\frac{1}{4}$ h.p. continuously, is illustrated in *Fig. 4*.

Independent Feed Screws

If the lathe is to be used for much milling, and in particular when the work must be mounted in the chuck, some means of independently driving the feed screws has to be devised. If it is not the user will always be obliged to adopt the somewhat tedious practice of turning them by hand.

So far as the leadscrew is concerned, perhaps the simplest method is to drive it through a train of gears from the lineshafting or countershaft. An example of such a drive may be seen in *Fig. 5*. This set up required an 80 to 1

Fig. 3. Overhead fitted to the Myford lathe driving a spotting drill

Fig. 5. Driving the leadscrew independently of the lathe

reduction gear box and some gearing to couple it to the leadscrew. If the drive needed to be reversed all that had to be done was to cross the primary driving belt. Later a low-voltage electrical motor replaced the line shaft drive. This motor has a reversing switch thus providing for a change in feed direction when desired.

Subsequently, the drive has been modified so that the motor unit with gears can be transferred from one lathe to the other in the shop.

One of the advantages of the electric motor drive is the ability to control the rate of feed by varying the speed of the motor itself. This can easily be arranged by the use of a simple variable resistance in series with the low-voltage electrical supply. Since the motors used take only some 2 amperes the physical size of this variable resistance is small.

In order to make the most of the lathe for milling purposes the cross-

slide also needs to be independently driven. It is seldom possible, without some elaboration, to arrange this by straightforward mechanical methods; on the other hand, by a combination of worm and epicyclic gearing, it is possible to build a neat driving unit that can be attached to the tail of the cross-slide. The reduction ratio is of the order 1,500 to 1 so the torque at the feed screw is large, thus permitting the use of a low power electric motor. Again, a reversing switch can be fitted to take care of the requirement for a change in the direction of feed, while the speed of the motor is controlled by a rheostat.

The epicyclic gearing is controlled by a lever-operated cam contracting a steel brake band on to the drum forming the housing of the annular gear. By adjusting the tension of this band the housing can be caused to stop and so relieve any excess load that may develop between the cutter and the work.

Fig. 4. A low-voltage miniature drive unit

LATHE OVERHEAD DRIVES

Fig. 6. The tailstock drilling spindle set up in the lathe

A High-speed Tailstock Drilling Spindle

When small drills have to be used in the lathe it is seldom that they can be run at a high enough speed, unless special steps are taken to do so. The rig illustrated in *Fig. 6*, was designed and made to allow the drilling of long axial holes in small pins and other components of a like nature, it is in effect a means of converting the tailstock of the lathe so that it becomes an electrically driven drilling machine operating in a horizontal plane. In the class of work for which it was designed it is essential that the drilled holes should run true and not run off-centre. Probably the best way to ensure that the holes do run true is to adopt the contra-rotating technique. In this method the work is rotated in the opposite direction to the drill. This has the effect of greatly increasing the cutting speed of the drill and appears to promote straight running.

The equipment seen in greater detail in the illustration *Fig. 7*, consists of a spindle fitted with a chuck that is passed through the hollow barrel of the tailstock and seated in the driving unit. Both the spindle and the driving unit are carried on a pair of ball races and are illustrated in *Fig. 8* at A and B respectively. The device was originally fitted to an old Winfield lathe having its tailstock modified to permit a more sensitive feed. It has now been transferred to a Myford ML 7 lathe incorporating the Cowell lever feed tailstock attachment.

The driving motor, mounted on a spigot behind the driving unit is carried on the bracket seen in the illustration *Fig. 9*. It came originally from an old vacuum cleaner and, after cleaning and oiling, was placed in the container illustrated. In order to control the motor a switch is built into the end plate at the opposite end to the driving pulley.

A High-speed Headstock Spindle

The requirements that led to the production of the tailstock drilling spindle led also to the design of a headstock that would rotate at a high speed. From time to time designs have appeared, but, so far as we are aware, most are fairly complex and few, if

Fig. 7. The tailstock drilling spindle

Fig. 10. The drilling spindle in use on the saddle of the Myford lathe

Fig. 8. The drilling head and spindle (A)

bly mounted at the tail of the mandrel.

Such an arrangement allows the inner spindle to be driven directly from the motor normally supplying power to the lathe. It also enables the self-act arrangements usually associated with the lathe to be used to the full. In point of fact six rates of feed are available, three in direct drive and three more with the back-gear engaged.

The inner spindle is adapted to take any, allow the lathe self act to be employed.

The success of the tailstock drilling spindle suggested that a complementary headstock spindle to the same design would adequately meet the case.

The headstock spindle, therefore, passes through the hollow mandrel of the lathe, is carried on a pair of deep-groove races placed in a housing attached to the mandrel nose, and is supported in a second bearing assembly A size collet chucks with a maximum holding capacity of a $\frac{1}{4}$ in., these are secured by means of a hollow draw spindle operated from the tail of the mandrel.

It is manifestly impossible to include here all the detailed drawings needed to produce the parts for any of the devices just described. But it is hoped that enough has been said to encourage and guide those who have a use for such equipment to design and make similar devices for themselves.

Fig. 9. The driving motor and bracket

CHAPTER 27

Soldering, Brazing and Case Hardening

SOFT soldering is a practice much in use by the amateur and in some fields by the professional, though the latter now tends to use other methods because of the increasingly exacting demands of industry when the jointing of metals is undertaken.

Solders and Fluxes
A hundred years ago the solders used would probably have been a single amalgam of lead and tin, in proportions almost entirely empirical, whilst the fluxes employed were resin when the work was exceptionally clean, or 'killed spirits' a preparation produced by the action of hydrochloric acid on zinc clippings, if a more vigorous fluxing action was needed.

Resin flux is non-corrosive but killed spirits is very much the reverse and its presence in the close confines of a small workshop can be most damaging to any machines and equipment that are installed in it.

Today, however, matters are much improved; not only are solders produced to meet almost every requirement, but the fluxes now employed, whilst thoroughly active, have for the most part been robbed of their highly corrosive properties. In addition many of the solders possess a core of the flux, and this renders them very easy to use. They are drawn as wires of varying gauges having lead and tin mixed in different proportions, each with an alternative melting point.

Methods of applying solder vary. In some cases a hot soldering iron, is used whilst in others solder in the form of rings or strips is placed on the work which is then heated by a blowpipe until the solder runs into the joint.

This is typically an industrial method and suggestions with practical examples are usually obtainable from any well known solder manufacturer.

An alternative method of applying the solder for 'sweating' as the term has it, is to make use of a solder paint. This is a mixture of very finely divided solder grains and an active flux, enabling the material to be used in the most economical manner. Only a very light coating needs to be applied to the work so positional disturbance of the parts to be formed is, for the most part, reduced to acceptable limits.

Heating the Work
Where it is to hand the domestic gas supply is a very convenient method of heating both the soldering iron and the work itself. While gas heaters similar to boiling rings but provided with a hood are obtainable, it is per-

Fig. 1. The self-blowing blowpipe

haps simpler, if the right equipment and gas are available, to make use of a laboratory tripod and Bunsen burner for the purpose. For heating small and delicate work the self-blowing blow-pipe illustrated in *Fig. 1* and in detail in *Figs. 2* and *3* has been found most effective. It is quite easy to make and has good flame control. Experiments, however, have shown that the torch is only suitable for use with coal gas.

Today the availability of bottled gas in its various forms, and the fact that many workers in the amateur field live outside areas where coal gas is on supply, is rapidly leading to a bottled gas monopoly that is scarcely likely to be disturbed in the foreseeable future.

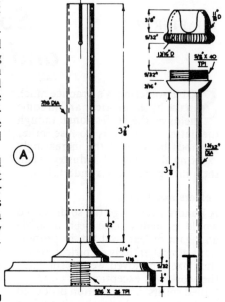

Fig. 2. Details of the stand

Fig. 3. Details of the self-blowing blow pipe

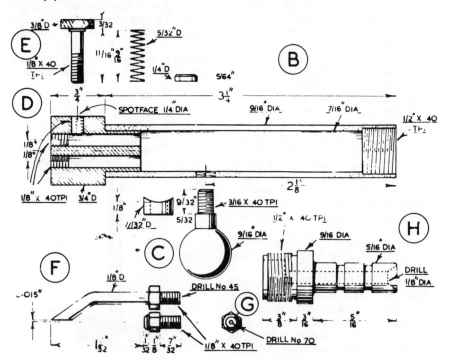

SOLDERING AND BRAZING

Moreover, the range of equipment available, the high calorific value of the gases themselves, and the convenience of being able to transport the gas-bottles and torches anywhere with the minimum of difficulty, is an asset that cannot be disputed.

For soldering purposes, therefore, a Bunsen burner specifically designed for use with bottled gas, a laboratory tripod and an ordinary soldering iron will furnish all that is needed.

Those workers who regularly undertake electrical wiring of one type or another will not need reminding of the advantages of the electric soldering iron. But it is, perhaps, not out of place to remind readers that quite large electrically heated irons are available, though some workers may object to the trailing cable that is inseparable from these devices.

Preparing the Work

The keynote of successful soldering is cleanliness. The work must therefore be free from grease, paint, rust or any metallic oxide deposits that will inhibit the flow of solder and prevent it from tinning the work surface. A small amount of dirt is certain to remain, if only in a chemical condition and it is the purpose of the flux to remove this. But the major part of the cleaning must be performed by the operator. Work surfaces that have been freshly machined need little more than degreasing with a suitable cleaner such as trichlorethylene to fit them for the soldering operation. But old work must be wire-brushed, scraped, filed or cleaned up with abrasive cloth as required in order to restore the surface to a condition in which the solder will 'take'.

Rusted parts are the most difficult to treat because, until they have been first chemically cleaned with a warm 10 per cent solution of phosphoric acid or some proprietary medium such as Jenolite, it will be difficult to assess if there is really any of the original metal left with which to effect a joint. If there is then the parts should be wire brushed and separately tinned before being sweated together. 'Bakers Fluid' may be suitable as a flux, but more active ones are available and workers would do well to consult solder manufacturers about them.

When preparing work for soldering and in particular where a sweating operation is contemplated, it is essential that the parts suffer no positional displacement during the process. This requirement may be met either by machining the parts so that they key together when assembled and cannot possibly move in relation to one another during the soldering operation or secure them in place using G-clamps for the purpose. In some instances a combination of both methods may be needed.

Hard Soldering

The process of hard soldering, or 'silver soldering' as it is sometimes called, is a somewhat similar operation to that just described. In this instance however, the temperature needed to carry out the work, some 700°C to 800°C, a dull red heat, may require something more than the heating we have previously considered.

For the most part hard solders have a silver base with copper and other metals in varying proportions. Perhaps the best known, and certainly most widely used is 'Easiflo' produced by Johnson-Matthey of London. This hard solder may be obtained as wire, strip or, very conveniently, as a paint where the metal is admixed with an active flux. The flux used is for the most part Boracic Acid in powder form with an activator added; for this reason it is advisable to make sure that work of a ferrous nature is not overheated or it may scale badly. All that is needed is a dull red heat when the silver solder will be seen to melt readily and to flow uniformly around the joint forming a fillet requiring the minimum of cleaning up, if any at all.

Equipment Needed for Hard Soldering and Brazing

We have already referred to the use of bottled gas in connection with soft soldering. This method of heating the work is eminently suitable when hard soldering is to be undertaken. As has already been said bottled gases have a high-calorific value, are clean, and the apparatus employed very controllable The two principal forms of burner suitable for brazing and silver soldering with bottled gases are depicted in the illustration *Fig. 4*.

The torch 'A' is really a form of Bunsen burner equipped with a handle and is very suitable for heating up large objects. It needs no air supply under pressure whilst the torch 'B' does require this. The second of the two torches illustrated gives a more concentrated and steady flame and so is the better for dealing with fine work.

With equipment such as this the objects to be brazed need to be surrounded by some refractory material in order to conserve the heat applied and to ensure that all parts of the work reach an even temperature. Formerly gas coke was used for the purpose but for many years now fire-clay cubes have been employed as they are quite inert and produce no fumes likely to damage the work.

In order to make the best use of this refractory material some form of hearth is essential. In *Figs. 5* and *6* two hearths that have been in our workshops for over 30 years are illustrated. The first of these fitments, originally intended for use with a paraffin blow-lamp, has been adapted to employ a coal-gas blow-pipe and is of commercial origin. The second, making use of a small lathe-side turner's cabinet, is equipped, as will be seen, for bottled gas. Both have a firebrick lining to the bottom and backs of their trays and each has plenty of refractory material for packing around the work.

A Rotary Blower

In a subsequent chapter the use of compressed air in the workshop will be discussed. Whilst it is possible to employ compressed air in connection with a bottled-gas blowpipe such as that illustrated in *Fig. 4* owing to their sensitivity it is better to make use of a rotary blower similar to that depicted in *Fig. 7*. The particular contrivance illustrated is a piece of home-made equipment and is of the eccentric vane type; that is to say its rotor is set off-centre in the stator so that the air drawn in through the induction

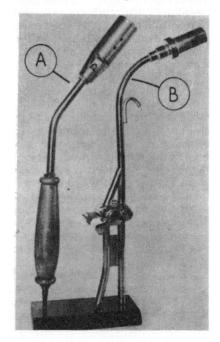

Fig. 4. Burners suitable for bottled gas

SOLDERING AND BRAZING

Fig. 5. Brazing hearth—commercial pattern

port is gradually squeezed into a reduced space between each pair of vanes in turn and is then expelled through the delivery port into an oil separator. The device is self-lubricating, oil being delivered through a sight-feed from the separator, which also acts as a reservoir, to the induction port through the small pipe seen in the illustration. The main bearings make use of ring lubricator and oil is supplied to them by means of an oil can through the aluminium cover seen over the bearing housing.

A blower of this kind supplies air at low pressure but in relatively large quantities and this is the condition most suited to the use of bottled gas.

Preparing the Work

The preparation of the work for brazing or silver soldering follows very much the procedure adopted for soft soldering. The work must be clean, freshly machined chemically cleaned work giving the best results. The parts to be joined must be a good fit, the maximum permissible slackness of such components as pipe unions being some 0·001 in. The parts are assembled together with the joint faces fully fluxed so that, when the work is heated, the brazing material will flow freely.

The use of silver solder paint, when applicable, is a great help since it provides the best means of thoroughly wetting the work surfaces, and, for the most part, enough solder to form a neat joint.

When brazing flux is heated and becomes mixed with oxides, a glassy substance is formed on the surface of the work. Naturally, this needs to be cleaned off. Hot water and scrubbing will usually do this successfully, but stubborn cases may require an acid pickle to remove the hardened flux.

Hardening of Cutting Tools

The cutting tools used in the amateurs workshop, certainly those he makes for himself, are derived from a material known as 'silver steel'. This is a steel containing sufficient carbon to enable it to be hardened, if heated and then quenched in a suitable coolant.

The tools are heated to a cherry red colour and then plunged into clean cold water, or iced water if a some-

Fig. 6. Brazing hearth for bottled gas

Fig. 7. Low pressure rotary blower

water to arrest the tempering operation.

When a number of parts have to be tempered to a uniform state use can be made of a sand bath. This consists of a metal box, filled with silver sand, provided with some means of heating such as a gas ring or bunsen burner. A typical set-up is illustrated in *Fig. 8*. To employ it the sand is first heated until a sample component, already buried in the sand, is seen on inspection to have taken on uniformly the tempering colour required. The heat is then removed from the sand box and the remainder of the components placed in the sand.

They will, after a suitable interval for soaking, all reach the same degree of tempering as the sample, but no greater, and can then be withdrawn

what greater hardness is required. The steel is then dead hard, a condition rendering it unfit for immediate use because of its lack of mechanical strength.

Accordingly the tool needs to be tempered until its brittleness has been removed whilst still leaving the cutting edge hard enough to withstand prolonged use.

In order to temper the tool it is first polished all over to ensure that the oxide film produced by the hardening process is completely removed. Heat is then applied well away from the cutting edge and maintained until a play of colours is seen to commence. The colours range from light straw through dark straw and blue to dark blue. The cutting edge of tools intended for cutting metal should not be a deeper colour than light straw though the shank, of a boring tool for example, will naturally be a deeper colour. Wood cutting tools usually have their edges tempered to a dark straw.

When the desired colours have been reached the tool is plunged into cold

Fig. 8. Set-up for sandbath tempering

Fig. 9. The electric muffle

and cooled either in air or, if wanted in a hurry, by being plunged in cold water.

High Speed Steel

The hardening and tempering of high-speed steel need not concern the amateur worker or even the small professional shop. As has been stated elsewhere fully treated round or rectangular section tool bits are available. These only require grinding to shape. In any case the equipment necessary for the treatment is more complex than would be warranted in the amateur workshop.

Case Hardening

Case hardening is the process whereby steel components have their outer surface layers converted into a material that may be hardened in the same way as tool steel, at the same time leaving the core of the work in the softened condition.

In this way the necessity for tempering is avoided and the skin or case may be left glass hard. Examples of case-hardened parts are those used in the production of motor cars where it is essential that, in addition to having a hard wearing surface, the components should have adequate mechanical strength.

We need not concern ourselves here with the precise metallic changes that take place. Suffice it to say that the case-hardening process increases the amount of carbon present in the steel and converts it from what is virtually a mild steel into a tool steel.

This condition is brought about by heating the work to between 800°C to 1,000°C in the presence of a car-burising agent. There is a number of proprietary substances on the market suitable for treating small components, but undoubtedly, if one is fortunate enough to obtain it, bone dust as used by the gun-makers provides the best medium. The work is packed in a cast-iron box, the cast boxes used for electrical wiring are excellent for the purpose, and is then raised to the required temperature and held there for a period depending on the depth of case desired.

Methods of heating small work vary from the simple use of the blowpipe and brazing hearth to the ideal, the electric muffle.

A small muffle, made by amateurs, is illustrated in *Fig. 9*. This device, controlled by a simmerstat as used in cookers, consists of a vitrosil quartz chamber, over which a heating element is wound. This chamber, as may be seen in the illustration *Fig. 10* is supported in a frame with end plates one of which is provided with a door having a refractory brick facing. The whole is enclosed by side plates and the interior filled with vermiculite as a heat insulator. The illustration shows an electric pyrometer in use with the muffle, but this is not essential. If, when case-hardening for the most part, the work is heated to a dull red colour, (gun makers call this 'worm red') the temperature reached will be correct and the work when chilled will take on the pleasing mottled appearance often associated with high-class work.

Case Hardening Procedure

We have already briefly referred to the basic methods to be used when case-hardening small parts. It is time now to consider the sequence of operations in greater detail.

1. The work must first be thoroughly degreased and cleaned to remove any traces of dirt that would otherwise spoil the mottled finish. Use trichlorethylene or carbon-tetrachloride, carrying out this part of the operation either in the open air or somewhere having plenty of fresh air available.

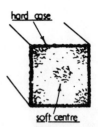

Fig. 11. Cross-section of material to show unhardened and hardened conditions

2. Next, the parts are packed in bone dust, or some form of animal charcoal, using cast-iron containers that will fit the muffle and allow the introduction of enough of the carburised material to ensure the parts are fully covered. This is important, for no air must get to the work whilst it is being carburised or the surface finish will suffer. Do not use 'bone meal' obtainable from horticultural suppliers. This contains a certain amount of sugar, a substance that will contribute nothing to the hardening operation, rather the reverse.

Fig. 10. The muffle—to show construction

3. Raise the work to temperature and hold at this level long enough to obtain the depth of case required. The time will vary to some extent on the steel used to make the parts, but a maximum depth of some 0·010 in. per hour is a permissible basis for calculation.

    It is a good plan to put into the box small pieces of the parent steel so that these can be case-hardened at the same time as the main components. These small pieces can then subsequently be broken and used as a check on the depth to which the hard case has penetrated.

    The depth of penetration can readily be ascertained from an inspection of the cross-section of the broken pieces.

    In *Fig. 11* the metallic cross-section presents an appearance crystalline in the centre with a pearly grey fine grained rim. It is this rim that is the hardened case.

4. The final operation in the hardening process is the plunging of the work into cold water. This should be done as quickly as possible in order to ensure that the air has no time to act on the red-hot work. The water should be as cold as possible and must, of course, be absolutely clean.

CHAPTER 28

Compressed Air in the Workshop

THERE are many uses for compressed air in the workshop, so the provision of a simple but efficient supply system would seem fully justified. Unfortunately commercially made equipment is expensive and the amount of use it is likely to get in the amateur shop would scarcely warrant the initial cost. However, much may be done by the amateur himself to provide adequate apparatus of his own making, indeed our own workshops are so equipped, with modified material obtained for the most part on the surplus market.

The keynote of any system is the compressor itself, and this should be capable of providing at least three cubic feet of free air per minute. Otherwise the demands of the various air-driven tools are not likely to be met, so their performance will naturally suffer. It is better to have available too much air rather than too little.

In our own shops a pair of compressors of the type illustrated in *Fig. 1* are used, both compressors are separately mounted on small two wheeled trolleys and each has its own air receiver also attached to the trolley.

In this way both compressors, which are of course independently driven, may either be used together or separately as the occasion demands. They are normally kept together in the shop but can be removed for duty elsewhere when needed.

Useful air-receivers for the purpose are the oxygen cylinders once supplied as aircraft equipment. For the most part these cylinders are made from stainless steel and so are proof against the corrosive action of the condensate inseparable from compressed-air systems.

Fig. 1.
The workshop compressor

COMPRESSED AIR IN THE WORKSHOP

Fig. 1A. A small compressor driven by a low voltage motor

Fig. 3. The reducing valve

Fig. 2. The separator

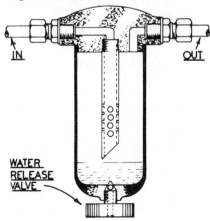

A useful maximum working pressure for the air line is 100 p.s.i. So any receiver should be hydraulically tested to at least 200 p.s.i.

While the air line itself can be of a permanent nature say of steam barrel with points for the quick-attachment of air hoses, probably one point in the shop and another say in an outside building such as the garage will be enough to satisfy most needs.

The hoses employed should be of adequate mechanical strength. While a burst here is scarcely calculated to result in a catastrophe, failure of the air hose when in use is always a nuisance.

Since, in any compressed air system a quantity of water vapour is formed, it is necessary to provide a filter or separator that will remove it. For many purposes the presence of water is not detrimental, provided of course that the quantity is not excessive, but for spray painting it is essential that the air should be dry and free from moisture.

The separator is a simple device consisting, as illustrated in *Fig. 2*, of a cylindrical container fitted with inlet and outlet connections and a quick release valve enabling the water to be drained off.

For all practical purposes a simple device of this nature will effectively de-water the workshop air supply, but must not, of course, be expected to deliver air dried to meet the conditions needed in some laboratory work.

The Reducing Valve

When using the workshop air supply in connection with brazing torches fed with propane or similar bottled gases a reducing valve of the type illustrated in *Fig. 3* is well nigh essential. Bottled gas torches are very sensitive to air pressure, and are almost impossible to control by any other means. There is a number of types of reducing valve, the simplest being that shown in *Fig. 4*

Fig. 5. Diagram of reducing valve to show principle of operation

depicting in section one made in the 'Duplex' workshops.

The principle that the reducing valve operates on is simple:

Compressed air is fed into the device through the valve A, *Fig. 5*, and impinges on the metal diaphragm B, thus immediately shutting off the supply. A compression spring C, however, is attached to the upper side of the diaphragm and has its tension made adjustable by means of the screw D. In this way the effect of air pressure on the diaphragm can be balanced against the spring tension itself. Therefore increasing this tension by screwing down the adjuster serves to raise the air pressure available at the outlet and to maintain it at the value required. This arrangement also enables the device to handle small fluctuations in the pressure of the incoming air supply.

A typical application of the reducing valve is that illustrated in *Fig. 6*. Here the valve is being used to control the air supply for an aerograph-type pen as used in the work of technical illustrators, an input air pressure of some 60 to 80 p.s.i. being regulated to 15 to 20 p.s.i. at the pen itself.

Fig. 6. A typical application of the reducing valve

Fig. 4. The reducing valve in section

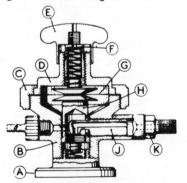

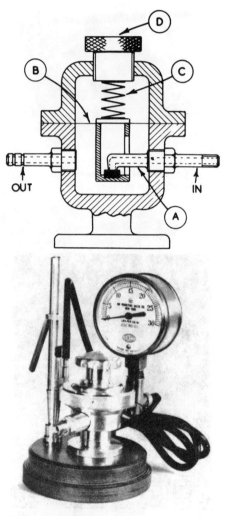

Uses of Compressed Air in the Workshop

Air is most commonly used in the workshop for cleaning down the machine tools after use. This is a matter upon which there are two schools of thought, and even Duplex between themselves have held divergent opinions.

Some contend that the air forces small particles of abrasive swarf into places they would not normally reach

COMPRESSED AIR IN THE WORKSHOP

Fig. 8. An air-driven handpiece with mounting for use in the lathe top slide

Fig. 7. A turbine-driven drilling attachment

by natural means; others say that modern machine tools are so well protected against the ingress of swarf that it could not possibly gain access even when driven along by the pressure of compressed air.

This latter view would seem to be borne out by industrial experience, for there seems no evidence to show that in the professional workshop at any rate, machine tools are having their useful life-span reduced or curtailed by the use of compressed air for cleaning-down purposes.

However, be that as it may, the user should be discreet or damage may eventually result; moderation here, as often elsewhere, would seem the best advice.

The gun used for the purpose is a simple piston-operated device, usually these days made in plastic material, attached to a hose, some 12 ft. in length, that is both strong and flexible. There seems little point these days in making a cleaning-down gun for oneself. Commercially produced devices are relatively cheap to buy and the time saved by **not** making them can doubtless be better spent. The same remarks probably hold true for the means of attachment used in connecting air hoses to the air line itself. For a very simple system a union and nut is all that is necessary. But manipulating them is a somewhat slow process and cannot be compared with the facility offered by the modern air-chuck, that enables a hose to be connected or disconnected in the matter of seconds. Those who feel that greater versatility will amply repay any slight extra cost may not grudge the few shillings needed to obtain it.

Blowpipes

In the amateur workshops the air supply can be used conveniently to feed the gas blowpipes needed in connection with the brazing hearth.

Where Town Gas is the fuel, nothing more elaborate than another tap is needed to control the air supply. But with bottled gas, as we have already seen, this will not do, the sensitivity of the apparatus to critical control of the air pressure makes a reducing valve essential.

Some Specialist Equipment Needing Air Supply

Those who have a supply of compressed air available may be interested in the two pieces of equipment illustrated in *Fig. 7* and *Fig. 8* respectively, the first illustration depicts a turbine driven drilling attachment capable of the high speeds needed to drill very small holes. The second illustration shows an air-driven handpiece suitable for use with small grinding wheels. Both these pieces of equipment were made in the 'Duplex' workshops in order to solve some machining problems posed by a government research department.

Spray Painting

Of all the uses that may be made of compressed air in the amateur workshop, perhaps the ability to carry out spray painting in some of its many forms is the most important.

Practically any pigment can be sprayed provided the correct type of gun is employed. For the amateur there are two basic forms, the gravity gun and the suction gun. In the first the paint is delivered to the nozzle by gravity whilst in the second it is raised from the container by suction. The gravity gun is illustrated in *Fig. 9* where the paint container can be seen **above** the gun as opposed to the suction gun where the pigment is placed **below** the nozzle as seen in the illustration *Fig. 10*.

The gravity fed gun is used for the

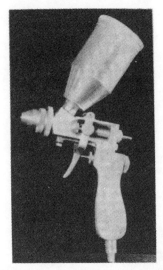

Fig. 9. A gravity-feed paint gun

more viscous fluids such as the heavier stoving enamels whilst the suction feed gun enables light fluids to be handled rather more simply.

Both guns employ compressed air to atomise the pigment and deliver it finely divided to the component being painted.

Fig. 10. The suction-feed paint gun

COMPRESSED AIR IN THE WORKSHOP

The gravity feed gun is provided with two valves, one, the upper valve seen in the illustration, controls the paint supply whilst the other, set in the handle of the gun, deals with the air supply.

The timing of the two valves is so arranged that air is turned on in advance of the paint supply. In this way spitting or blobbing of the paint is avoided, the more so since the paint valve has a tapered needle enabling the pigment itself to be supplied in gradually increasing quantities up to a pre-set maximum amount.

Per contra, the suction fed gun, has but one valve and this controls the air supply. The quantity of paint supplied is adjusted by the position of the combining cone in relation to the air nozzle seen in the illustration *Fig. 11* where the details of the various parts are depicted diagrammatically.

The action of the gun is simple. Air is supplied under the control of a valve located at the top of the handle and operated by thumb pressure. The air then passes to the air nozzle where it issues to atmosphere through the combining cone thus causing a depression in the body of the gun itself. Paint is then drawn from the container and delivered to the combining cone where it is finely atomised before being sprayed on the work.

In order that the device shall operate properly it is essential that the air nozzle is located centrally in the combining cone. To ensure this a centring device consisting of a small length of brass gearing, made a good sliding fit in the body of the gun and bored concentrically, is sweated or brazed to the air tube itself.

In this way the air is supplied correctly to the combining cone whilst the pigment, passing along the teeth of the centring device, also finds its way there. An illustration showing the parts in question is given in *Fig. 12*.

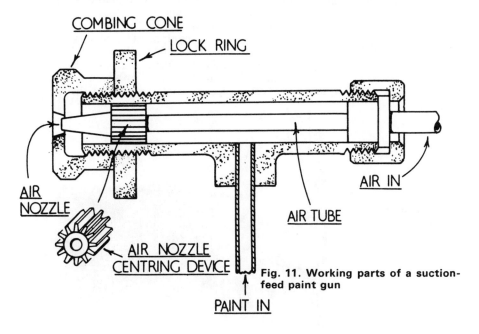

Fig. 11. Working parts of a suction-feed paint gun

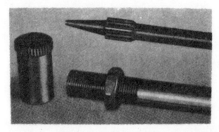

Fig. 12. The combining cone and air jet centring device

We have already seen that almost any form of paint can be applied by means of compressed air, using the gravity gun for heavy or viscous liquids whilst the suction fed equipment is used to handle light fluids such as well-thinned cellulose or synthetic pigments. The amateur, for the most part, will content himself with this class of paint and it is worth noting, in this connection, that many well-thinned coats are preferable to a few comparatively thick ones, the same remarks applying equally to oil based paints where these are used. In this way 'runs' or 'tears' can be avoided.

Stove Enamelling

This is a process that enables the user to apply a harder and more durable type of paint, and so is very suitable for finishing some of the smaller items of equipment or mechanical devices that have been made in the workshop. Stove enamels are made to give a hard glossy surface, or, alternatively, to provide a wrinkled or 'crackle' finish to the work. These latter enamels are somewhat viscous and so need a gravity feed gun to apply them. They must be used as supplied and cannot be diluted, moreover, as they have a somewhat short shelf life, they need to be employed as fresh as possible or they will not wrinkle uniformly.

As their name implies these enamels need to be 'stoved' in an oven in order to cure them. This is work that **can** be carried out in the domestic cooking stove but is unlikely to find much favour in the sight of the domestic authorities. Those readers who are contemplating stove enamelling operations may consider the purchase of a suitable oven worthwhile especially as these are often obtainable at auction very cheaply.

Illustrated in *Fig. 13* the complete set of equipment needed to stove enamel satisfactorily the small details and components that need this treatment. It will be observed that both types of paint gun are used, the gravity

Fig. 13. Complete equipment for stove enamelling

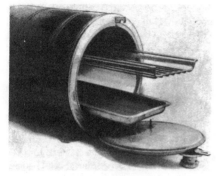

Fig. 14. The interior of the enamelling oven

feed gun for the heavy enamels that cannot be diluted and the suction or injector gun for use with small parts needing several coats of thin paint.

The enamelling oven is of some interest since it is almost ideal for this type of work. It is run directly from the domestic electrical supply under the control of a simmerstat switch enabling the temperature in the oven to be adjusted and maintained at very close limits. The oven is provided with a thermometer so that the operating temperature can be seen at a glance.

The interior is almost perfect for the purpose in hand in that, as may be seen from the illustration *Fig. 14*, there is a grating from which small parts may be hung and a tray to catch any surplus pigment that may inadvertently drip from the work during the stoving process.

As to the correct temperature for successful stove-enamelling this is a matter related directly to the type of pigment to be used; queries concerning this matter should therefore be addressed directly to the paint makers who are always ready to give advice. One should perhaps, state, however, that the temperature range is from 100°C to 200°C and that these comparatively low temperatures are, no doubt, one of the reasons that an oven bought some 30 years ago for 10 shillings is still in operation.

Preparing the Work

Parts that have been machined will need but little preparation, but castings and all rough surfaces should be made as smooth as possible, using a filler followed by a surfacer-primer in the case of cellulose or synthetic paints.

Work for stove enamelling may sometimes need a filler and the paint people should be consulted about this aspect of the process.

However, before any pigment of whatever type is applied, the work must be thoroughly degreased using a solvent such as tri-chlorethylene. It is good practice to make sure that the work is thoroughly dry before painting otherwise occluded solvent or grease may be driven out during the stoving operation; in this event the work will undoubtedly be spoilt and will have to be stripped before being repainted. A light stoving in the oven to dry the work before painting will avoid this trouble.

According to the class of work and to the material from which the parts are made so the use of priming coats will depend. It is really not possible to generalise about this so it is best to again ask the paint manufacturer for his advice.

However, it may be stated that of the metals commonly in use, mild steel and cast iron pose no problems, brass is somewhat unpredictable whilst the aluminium alloys are the most difficult needing an etch-primer to provide a good key for the paint.

When using a paint gun, unless large areas are being treated, it is best to apply the pigment in short bursts. The air pressure used, whilst not critical, will vary from 30 to 40 p.s.i. for well-thinned paints to some 60 to 80 p.s.i. when heavy enamels are being applied. At all times it is essential that the air supplied to the gun is free from condensate or the work surface will be spoilt.

It will be appreciated that, when painting an object by the spray process, it is not possible to control the limits covered by the pigment unless certain precautions are taken. If the work is to be treated all over with a single colour then these precautions will not be necessary, but when two or more tints are used, or some parts of the work need to be left bare, then the work must be masked as it is termed. For this purpose masking tape is used. This is a highly adhesive paper tape, available in a number of widths from ½ in. wide to as much as 4 in. wide. Small objects are usually masked by applying the tape directly to them whilst large surfaces such as the windows of motor cars for example, are masked with newspaper held in place by tape. If left on for any length of time the tape, on removal from the work, tends to leave behind particles of its adhesive compound. Petrol will dissolve this unwanted material if applied with a piece of rag, but it is inadvisable to use any other solvent or the work surface may be spoilt.

When painting small or medium sized components a rotary table to support the work is an almost essential accessory. A table of this type enables the work to be rotated to face the gun step by step, or, if made heavy enough, to revolve almost continuously at a slow rate during the painting operation. The table shown in the illustration *Fig. 10*, has been in use many years. It is employed as illustrated for step-by-step operation and has an old heavy flywheel placed on it when slow almost continuous rotation is needed. Rotating tables can be improvised in many ways, but it is important to make sure that the bearing used is as free from friction as possible otherwise the almost continuous rotation requirement will not be met.

Fig. 15A. Rotary paint table

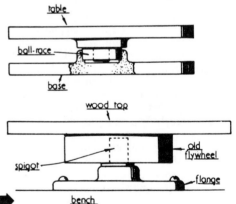

Fig. 15. Rotary paint tables

CHAPTER 29

Some Additional Machine Tools

Hacksaw Machines

Although the hand hacksaw is essential for occasional use in the workshop the machine hacksaw has many advantages. Not only is much labour saved, but the accurate cutting means less waste of material and leaves cut surfaces that need little after-work for the final finishing.

Moreover, the accurate guidance of the saw blade in a straight line saves blade breakage and protects against wear of the set of the saw teeth. The machine hacksaw illustrated in *Fig. 1*, made by Messrs. Cowell, is a robust and accurate tool, capable of dealing with work measuring 2 × 2 in.

Relief of the cutting pressure on the return stroke is provided for in accordance with the practice adopted in large commercial machines. To save expenditure the machine can be supplied in the form of a set of partly machined castings, which enable the hacksaw to be completed in a 3½ in. lathe.

A full set of working drawings is included. If required, a switch can be provided that automatically stops the electric driving motor on completion of the sawing operation.

The second machine, illustrated in *Figs. 2* and *3*, was designed and made in the workshop nearly 20 years ago, since then it has given every satisfaction and no alterations have been needed. Before deciding on the design and embarking on the construction, a test rig was set up to determine the correct rate of stroke and cutting pressure. In addition, wear of the saw teeth was investigated when no provision was made for relieving the cutting pressure on the return stroke.

So little wear was found on the saw

Fig. 1. The Cowell machine hacksaw

after a lengthy trial that it was decided to dispense with this added complication.

In fact, over this long period of use, renewal of the saw blade has been a rare occurrence, in spite of the large amount of work undertaken in making other machines and workshop tools.

The saw operates at 90 strokes a minute, and the primary drive from the ¼ h.p. electric motor is by a V-belt to a countershaft; from there the drive to the crankshaft by means of a connecting rod, fitted at either end with a ball bearing. The saw frame slides on a pivoted carrier-bar, and provision is made for taking up wear and ensuring adequate lubrication. The height of the pivot bar can be adjusted to set the saw blade for level

Fig. 2. The workshop hacksaw

Fig. 4. The workshop jig saw

Fig. 3. Showing the two-stage drive

SOME ADDITIONAL MACHINE TOOLS

cutting, as may be required in various machining operations.

Eclipse 12 in., high-speed steel, saw blades are used. These, when divided by the grinder cut-off disc wheel, make two blades for the machine. The hole already in the blade provides for one fixing to the saw frame and, at the other end, two notches are ground for clamping the blade in the tensioning device. A Myford machine vice of $1\frac{5}{8}$ in. capacity is pivoted to the steel table of the machine so as to allow for angular cutting. The cutting pressure is exerted, and varied as necessary, by a weight which slides on a rod attached to the bar carrying the saw frame. An

Fig. 5. The machine mechanism

automatic catch is provided for holding the saw in the raised position while setting the work in place in the vice.

The catch is released by a press-button at the front of the machine. An automatic switch, mounted on the machine table and fitted with a reset button, stops the motor when the saw reaches the end of its downward travel. This allows the machine to be left unattended during a protracted cutting operation and, later, the material will be found severed and the machine stopped.

Those wishing to build this useful machine can obtain a full set of working drawings from Messrs. Model and Allied Publications, the publishers of *Model Engineer*.

The Jig Saw

The machine illustrated in *Fig. 5*, which was designed and built in the workshop, has proved capable of carrying out a variety of work on metals, plastics and wood with a high degree of accuracy when the saw is following either a curved or straight path. As shown in the illustration *Fig. 5*, the drive from the $\frac{1}{4}$ h.p. motor is by V-belt to the crankshaft of the machine which drives a short, ball bearing, connecting rod that is coupled to a rocking lever, pivoted to a swing-link.

In this way, a reciprocating motion is imparted to the cross head of the spindle that carries the lower end of the saw blade.

Fig. 6. The ripping fence

Fig. 7. The mitring fence

At its upper end, the saw blade is attached to a spring-loaded piston rod working in a cylinder. The upward movement of the piston causes air to be blown through a nozzle for the purpose of clearing the saw dust. Eclipse jig saw blades, 6 in. in length between the driving pins, are supplied in a variety of widths and tooth pitches to meet all ordinary requirements for sawing both wood and metals. For cutting strips of material to width or for cross-cutting in length, the fence *Fig. 6*, is clamped to the machine table in the correct position.

As shown in the illustrations, an adjustable pressure-foot is attached to the machine to retain the work in contact with the machine table while the saw blade is rising.

Mitre-cutting is carried out by using the fence illustrated in *Fig. 7*. The blade of the fence carries an adjustable stop at one end and at its other end is a clamping device that holds the work in place and so ensures that the corresponding sides of a frame are cut to exactly equal length.

Fig. 8. A mitred camera frame

Fig. 9. Examples of work done in the machine

Fig 10. The workshop circular saw

SOME ADDITIONAL MACHINE TOOLS

Fig. 11. The saw drive

The frame, forming part of a camera, was made in this way, and *Fig. 8* shows that the joints, straight from the saw, have been accurately cut and have needed no hand-fitting.

Circular cutting of either discs or holes is readily carried out in the machine. For this work, a coned centre, on which the material rotates, is attached to the spindle carrying the blower nozzle. This form of mounting enables the radius of the cut circle to be adjusted as required. A narrow saw blade should be used to negotiate the circular path travelled. The circular hole in the stand for a voltmeter shown in *Fig. 9* was cut in this way. Sheet metal or metal strip can be cut in the machine when a suitable saw blade is fitted. To illustrate the capacity of the jig-saw machine, the lathe fitting also shown in *Fig. 9*, was slit through to the bore of this mild steel part, which measures 1½ in. in height.

The Circular Saw

The workshop circular saw, illustrated in *Figs. 10* and *11*, is mostly

Fig. 13. Work samples from the saw

used for cutting plastic materials and wooden parts. The drive is by V-belt from the ¼ h.p. electric motor direct to the saw spindle. The 4 in. high-speed steel metal saw which is used for cutting plastics leaves a good finish on the work and subsequent rubbing with liquid metal polish establishes a high finish. A wood saw is also available, but the metal saw is often used for small work of this kind. A slow driving speed is purposely used to avoid overheating when cutting plastic material, but the speed can be increased, if required, by changing over the two belt pulleys, substitution which is allowed for in the design.

The saw spindle, which is lapped on its two bearing surfaces, is carried in an iron casting, *Fig. 12*. This casting was trued up in the shaping machine before the bearing bores were machined in the lathe. The bearings were finally lapped to a close running fit for the saw spindle. Plain bearings have the advantage of quiet running at high speeds and, in the present machine, their diameter, which is smaller than

Fig. 12. The spindle bearing casting

that of ball bearings, also allows a saw of smaller size to be used.

A detachable swarf-box is clamped to the underside of the machine table to catch the saw dust and prevent it reaching the bearings.

The fence for ripping and cross-cutting and also the mitring fence are of the same design as those described for use with the jig saw.

With the exercise of a little ingenuity, and by using suitable packing strips, in conjunction with the machine fence, little difficulty has been found in cutting grooves, tongues and rebates such as those illustrated in *Fig. 13*.

CHAPTER 30

Back Tool Post

AS its name implies the back toolpost is a support for a turning tool capable of being secured to the rear of the lathe cross-slide opposite to the more normal top slide.

By this means the lathe operator has at his command an additional tool that may be brought into use without disturbing any equipment already secured to the top-slide.

However, the use of the back toolpost need not be confined to mounting a parting tool only. Those turners who regularly machine brass, and in particular the so-called 'screwing' quality, will not need to be reminded of its capabilities for distributing swarf far and wide. A rear mounted turning tool largely eliminates this nuisance since the swarf is discharged downwards.

Boring tools may also be used, but for the most part these need a special mount because the toolposts commonly supplied are suitable only for tools applied at right angles to the work and not along the centre line.

One such toolpost is that fitted to the Myford lathe. As may be seen from the illustration *Fig. 1*.

The toolpost consists of a body fitted with a hardened tilting boat

Fig. 1. The Myford back toolpost

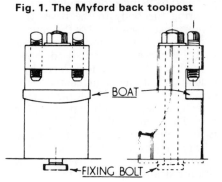

◀ Fig. 3. Prototype 2-tool back toolpost

Fig. 2. Alternative toolpost

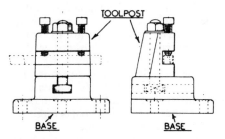

Fig. 4. Back toolposts—types D and M

enabling the tool itself to be adjusted for height, though of course with some adverse effect on tool rake. Errors in this matter, however, may be corrected by interposing packing where possible between the tool and the boat. The body is secured to the cross slide by a single bolt. An alternative toolpost is seen in the illustration *Fig. 2*.

This toolpost comprises a base that may be bolted to the lathe cross-slide and a toolpost that is attached to the base. The toolpost may be swung through 360 degrees thus facilitating the mounting of any form of tool including boring tools. In point of fact this toolpost is one of a pair, the other being designed to take the place of the top slide for certain operations.

The rear toolpost does not need to be confined to the mounting of a single tool. If a simple turret is provided, two or more tools may be used at the back of the lathe. When making a number of repetitive parts there are two operations that recur; they are chamfering and parting off. Both operations are well suited to rear-mounted and inverted tools and can most conveniently be performed with a turret enabling the requisite tool to be brought into play immediately.

The prototype back toolpost, illustrated in *Fig. 3*, is of built-up construction preparatory to making it from an iron casting in conformity with usual engineering practice. A casting ensures rigidity as well as materially simplifying manufacturing possibilities.

In the prototype rigidity was attained by using a long central screw securing the body to the base; this screw is extended to carry the clamping lever locking the tool turret to the body. When a casting is used, however, the central screw is replaced by a stud and the attachment secured to the cross slide by two T-slot tools located directly under the cutting point of the tool. The location of these T-slot bolts is an essential feature of the design. The placing of them directly under the toolpoint is intended to overcome the tipping strain imposed on the attach-

BACK TOOL POST

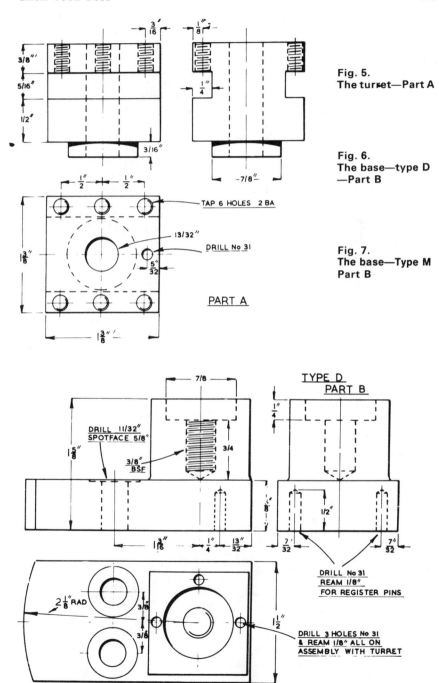

Fig. 5. The turret—Part A

Fig. 6. The base—type D —Part B

Fig. 7. The base—Type M Part B

ment when in operation, so it follows that, when the back toolpost is secured to the cross slide, care must be taken to see that it is placed in such a way that this design feature is satisfied.

The Toolpost

The two forms of toolpost are depicted in the illustration *Fig. 4*. For the sake of clarity they have been dubbed type 'D' and type 'M' respectively, type 'D' being suitable for the Drummond lathe whilst type 'M' is the form applicable to the Myford ML 7. This differentiation is maintained throughout the following description and is applied to all relevant drawings and instructions.

As will be apparent from the illustrations of the bases, *Figs. 6* and *7* (page 243 and below) the main difference, apart from some dimensional alternatives, is the method used to secure the two toolposts to their respective cross slides. The Myford form has a long central bolt passing through both the base and the turret together, with a second and shorter bolt holding the toe of the base casting to the lathe cross-slide. Each of these bolts normally engage a separate T-slot in the cross-slide. In the case of the Drummond toolpost, however, there is no central bolt, the base being secured by two bolts passing through its toe. The turret itself is then clamped to the base by means of the stud seen in dotted outline. Whilst this method of securing the device may seem less rigid than that used with the type 'M' no lack of stability has, in fact, been found in practice.

The turrets for both types of back toolpost are designed to accept the standard ¼ in. square high-speed steel toolbits now generally available. Accordingly the dimensions of the tool housings are arranged to suit

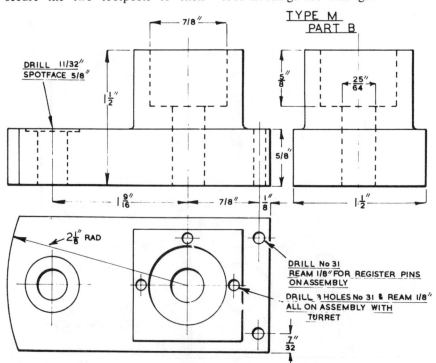

BACK TOOL POST

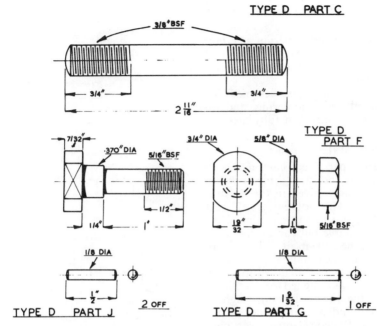

Fig. 8. Fixing bolts—type D

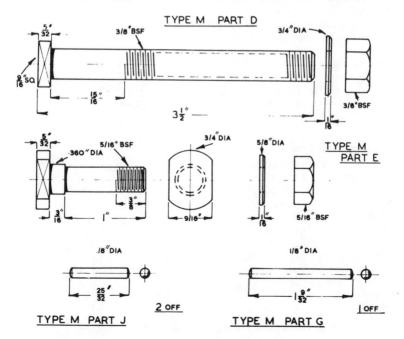

Fig. 9. Fixing bolts—type M

Fig. 10. Locking lever—part H

these toolbits and to permit a certain amount of adjustment for height, if necessary, when the tool is re-sharpened.

The two tools are designed for parting-off and chamfering, and we shall be discussing them fully later.

The Parts of the Toolpost

The turret, part A is common to both toolposts D and M. It is machined to the dimensions given in the illustration *Fig. 5*. The tool seatings are best formed in a shaping machine when one is available but can also be end-milled in the lathe itself with the work mounted on the top slide.

The base part B detailed in *Fig. 6* and *Fig. 7* was designed for machining in the lathe, the work being caught in the 4-jaw independent chuck. Castings for both main elements were once available and it is confidently hoped that this supply will be renewed.

The fixing bolts, nuts washers, and register pins applicable to both types of toolpost are detailed in the illustrations *Fig. 8* and *Fig. 9*. While the locking lever, part H is illustrated dimensionally in *Fig. 10*. This part is, of course, common to both types of toolpost.

Tools for the Back Toolpost

In all probability the finished toolpost will find a permanent location at the rear of the cross slide. In this event it will be used to house those tools that are employed in the last stages of machining. The tools in question are the parting and chamfering tools illustrated in *Fig. 11* and *Fig. 12*. They are conveniently made from high-speed steel bits, ground to the forms depicted in the illustrations.

Fig. 13. Eclipse blades

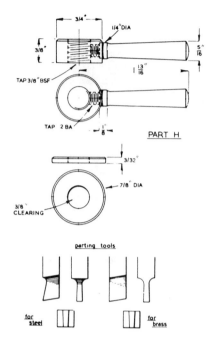

PART H

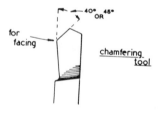

Fig. 11. The parting tool

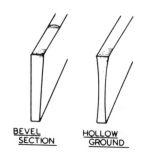

Fig. 12. The chamfering tool

BACK TOOL POST

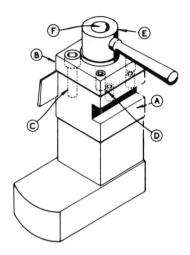

Fig. 14. Turret for Eclipse parting tools

Mounting 'Eclipse' Parting Tool Blades

Of the many additional tools that may be used in the back toolpost perhaps the most important and useful are the narrow 'Eclipse' parting tool blades. As may be seen from an inspection of the illustration *Fig. 13* these blades are obtainable in either straight-sided or hollow-ground form and that, in order to provide clearance in the work the blade is tapered in section. This taper has an included angle of approximately 3 degrees. It follows, therefore that any mounting for the blades must provide means of ensuring that the designed clearance is maintained.

For those who possess the Myford back toolpost only this requirement is satisfied by using the special tool holder provided by Messrs. 'Eclipse' themselves. This has the seating for the blade machined to the correct clearance angle and is provided with a clamping bolt to secure the tool in place. We regularly use the 'Eclipse' parting tool mounted in the Myford back toolpost and have found it in every way satisfactory especially when machining stainless steel that requires both rigidity and adequate clearance for good practical results.

Adapting 'Eclipse' Parting Tools to the 2-tool Back Toolpost

Those who wish to fit 'Eclipse' parting tools to the 2-tool back toolpost we have been describing will need to examine the illustration *Fig. 14* and *14A* where the details of the special turret to carry these tools are given. It will be seen that the turret (A) is provided with a cap (B) carrying two levelling screws (D) and that through this cap two blade tension screws (C) are passed clamping the blade in place. In addition two blade position screws threaded 6 BA are set in the side of the turret to ensure that the blade is correctly mounted to give the necessary clearance in work.

On the opposite side of the turret a

Fig. 14A. Turret for Eclipse parting tools

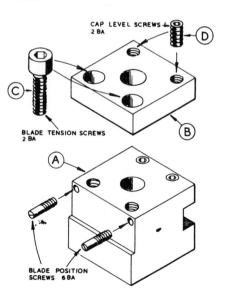

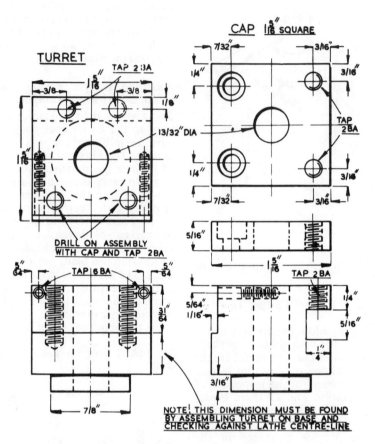

Fig. 15. Details of the turret

housing is formed for the combined chamfering and facing tool which is secured by a pair of allen grub screws in conformity with the provisions for the original 2-tool turret.

The blade position screws previously referred to were designed to accommodate an 'Eclipse' parting tool and set it correctly upright in a seating formed by an end-milling operation carried out in the lathe since for the majority of readers this was held to be the most convenient way of producing the seatings.

Those with shaping machine facilities however, will be able to machine the seating at the correct angle of 3 degrees.

CHAPTER 31

Reamers

REAMERS are used for opening out holes that have already been drilled or machined somewhat undersize. As an example, the fitting of a pin into a bush set in a particular component is usually performed with a reamer taking out only a few thousandths of an inch of material.

This is work that requires the use of the parallel reamer depicted in the illustration *Fig. 1*. As its name implies the tool is parallel for the greater part of its length, but has a very short taper lead to enable it to enter the hole being sized. It is intended for hand operation when caught in a suitable wrench, but is not designed for the removal of large amounts of metal. It may be used also in the lathe held in the tailstock chuck. The straight-fluted form shown in the illustration tend to reproduce their pattern on the work and are now generally replaced by reamers having a helical form since these do not exhibit the same tendency.

'Packing' the Reamer

A parallel reamer can only form a hole of the size to which it was originally ground unless the practice of 'packing' the tool is adopted. This is a somewhat haphazard procedure involving the use of paper or thin card placed over one or more of the blades in order to increase the effective diameter of the reamer as a whole. The same result can also be obtained by placing a length of round material into one of the flutes of the reamer.

Adjustable Reamers

In order to offset these difficulties two forms of adjustable reamer were introduced. The first of these, illustrated in *Fig. 2*, was of one-piece construction split down three or more of its flutes and drilled and tapped axially to accept the expander screw used for adjustment purposes. As might be expected their range of expansion was somewhat limited, only a few thousandths of an inch being possible, and they were also somewhat expensive to produce; so to provide greater coverage the reamer seen in *Fig. 3* was developed.

This form of reamer has a body slotted to accept its detachable blades. The blade seatings are tapered so that, when the blades themselves are moved axially by means of the nuts at each end of the body, the effective diameter of the reamer can be expanded or contracted at will.

When using reamers of this type great care must be taken to ensure that they are not asked to cut heavily, especially in the smaller sizes likely to be of interest to the amateur, for

Fig. 1. Parallel Reamer

Fig. 2. Adjustable reamer tapped axially for expander screw.

too heavy a loading may irretrievably damage them.

The range available is considerable and the coverage from any given size sufficient to provide overlapping between the sizes.

Machine Reamers

The reamer illustrated in *Fig. 4* is intended for use in machines. In the small workshop this will imply the lathe and possibly the drilling machine. In all cases machine reamers must be accurately mounted or they will tend to cut oversize. As is common with many fixed reamers when new, the machine reamer is usually made somewhat oversize to allow for resharpening. Before making use, then, of any non-adjustable parallel reamer it should be measured with a micrometer to establish its exact size.

Taper Reamers

For the most part reamers of the type illustrated in *Fig. 5* are employed for the purpose of sizing or dressing the Morse Taper seatings of machine tool spindles. This is an operation following on directly from the boring of the taper seating when the reamer is held in the tailstock chuck. Light cuts only are taken with the reamer and the work well lubricated to ensure that the work surface remains smooth.

When dressing existing tapers that may have suffered slight damage, the reamer is held in the hand and turned with a suitable wrench.

Taper Pin Reamers

If parts are to be held securely by means of a taper pin, not only must the parts themselves be a good fit but the seating for the taper pin that holds them must be correctly formed or the parts may start to 'work' and become loose.

To make certain that the pin seating is correctly formed, reamers ground to the correct taper are used. Since in the smaller sizes they are very delicate, for the most part these tools are used by hand, taking light cuts and cleaning the reamer of swarf by withdrawing it frequently and oiling it before replacement.

Fig. 4. Machine reamer

Fig. 3. Adjustable reamer slotted to accept detachable blades

Taper pins have a taper of ¼ in. to the foot, so the reamers are ground to suit. An example of a taper pin reamer is illustrated in *Fig. 6*.

The five-sided broach seen in *Fig. 7* is sometimes employed to seat small taper pins. But its rate of stock removal is slow and its action that of rubbing rather than cutting. Moreover, a broach tends to wander out of alignment if undue pressure is applied to it during the reaming operation. Nevertheless, used with care, broaches form a useful and comparatively cheap means of obtaining satisfactory pin seatings for the smaller sized taper pins.

In commercial practice taper pin reamers are used in quite large sizes, but in the small or amateur workshop the need will, for the most part, only be for the smaller reamers in the range, which has been designed so that the reamers themselves form a series, each one overlapping the next. The table below gives the salient information about those sizes likely to be of service in the small workshop.

Nominal diameter in.	Diameter at large end in.	Diameter at small end in.	Suitable for taper pins in.
1/16	0·064	0·0432	1/16
5/64	0·080	0·0592	5/64
3/32	0·095	0·0690	3/32
7/64	0·111	0·0798	7/64
1/8	0·127	0·905	1/8
9/64	0·143	0·1039	9/64

Fig. 5. Morse taper reamer

Some Notes on Reaming

In order to avoid overloading a parallel reamer, and to ensure a smooth finish to the hole, it is important to ensure that only the correct amount of material is left for removal.

Manufacturers recommend that this should be from 0·006 in. to 0·010 in. for reamers up to ½ in. in diameter and from 0·010 in. to 0·015 in. for reamers from ½ in. to 1 in. diameter.

Mention has already been made that new parallel reamers are purposely made somewhat oversize. The figures of tolerance in this respect are generally in accordance with the following table:

Reamer diameter in.	High in.	Low in.
Under 0·3 in.	+0·0006″	+0·0003″
0·3 and under 0·6 in.	+0·0008″	+0·0004″
0·6 and under 1·0 in.	+0·001″	+0·0005″

Expansion reamers of the type illustrated in *Fig. 2* have already been mentioned as having a limited range of expansion.

The small amount of expansion permissible is not always appreciated and often results in permanent damage to the tool. If the limits given in the attached table are followed no damage is likely to be caused:

Reamer diameter in.	Permissible expansion in.
¼ to 15/32 in.	0·005 in.
½ to 31/32 in.	0·008 in.
1 to 1 23/32 in.	0·010 in.
1¾ to 2 in.	0·012 in.

Finally a few words of caution when storing and using reamers. As it is of the utmost importance to retain their keeness reamers should always be

Fig. 6. Taper pin reamer

Fig. 7. Five-sided broach

kept either in stands similar to those used for drills or in drawers or cases where they can lie horizontally and are separated from one another by suitable partitions. Nothing is calculated to harm their usefulness more than to allow reamers to be jumbled up together in a drawer with their cutting edges in contact. It is also essential to see that rust cannot damage them. For this reason, when storing reamers, they should not be laid on any substance likely to absorb moisture. For this reason stands are perhaps preferable to drawers, unless some form of rust inhibitor can be introduced to overcome the trouble.

When using reamers they should always be turned in the direction of cut and never reversed on withdrawal, for their cutting edges are liable to be dulled if they are turned backwards. Practically all reaming operations can, with advantage, be undertaken with a lubricant of one form or another. The only materials to be reamed dry are cast iron, bakelite and the magnesium alloys. For the rest, apart from assisting the production of a smooth finish, particularly when reaming in a machine, the lubricant, if applied copiously helps to cool the work. The attached table, based on information received from the British Steel Corporation as well as from practical experience, will, perhaps form a guide when using high-speed steel reamers:

Material	Lubricant
Steel up to 80 t.p.s.i.	Sulphurised oil
Stainless Steel	Sulphurised oil
Malleable iron	Mineral oil: soluble oil or soda water
Aluminium alloys	Soluble oil: Paraffin and lard oil or paraffin
Brass	Soluble oil or paraffin and lard oil
Copper	Lard oil: paraffin sulphurised oil or soluble oil
Phosphor bronze	Lard oil or soluble oil
Plastics	Soluble oil or soapy water

Index

A
'ALBRECHT' Chuck, 37
Angular Grinding Rest. 75—76
Angle Plates, 188—189
Air, compressed, 226—232
Additional Machine Tools, 236—240
Adjustable Reamers, 249—250

B
Benches, 6
Bench Grinder, 16—18
Belt Drives, 41—46
Belts, Round, 41—45
Belt Fasteners, 41—42, 45—46
Belt Tensioning Devices, 46
Boring Tool, 71—72
Back Tool Post, 73—74
Box, Clapper, 50
Bell Chuck, 57
Boring Work in the Lathe, 86—88
Boring Tool with detachable cutters, 88
Boring Bar for small tools, 89
Back Facing, 90—91
Boring Tool Holders, Two Simple, 92
Boring Work on the Saddle, 92—93
Boring Bar, 93—95
Buttons, Toolmakers, 107—109
Blocks, 'V', 190
Blowpipe, Self-blowing 217—218
Back Toolpost, 241—248
Back Toolpost, Tools for, 246—247
Broach, 251

C
Cable Support, 10
Cooling of Electric Motors, 14
Centres, 22—24
Chip Trays, 30—31
Chucks, 36—38, 57—65
Chucks, 'Albrecht', 37
Chucks, 'Jacobs', 37
Chucks, securing drill, 36—37
Clapper Box, 50
Chucks, Bell, 57
Chucks, 4-Jaw Independent, 57—58
Chucks, Self-centring, 58—59
Chucks, Collet, 59—61
Chucks, Keyless, 61—62
Chucks, care of, 63—64
Chuck Brace, 64—65
Collet Chucks, 59—61
Chucks, Drilling Machine, 36—38
Cables, Electric, 9—10
Centres, Mounting Work Between, 84

Cutting Internal Keyways, 95
Circular Saws, 114—116
Chucks, Drill, 142
Countersinking, 155—156
Counterboring, 156—157
Centring Device, 154
Combination Squares, 193
Centre Punches, 194—195
Case Hardening, 223—225
Circumferential Dividing, 121—123
Compound Dividing, 123—124
Change Wheel Mounting, Lathe Mandrel, 125—126
Cutting Racks, 136—138
Cross-Drilling, 150—153
Countersinking, 153
Counterboring, 153
Colleted Die Holder, 161
Change Wheel Gearing 173—174
Change Wheel Gearing, Proving, 175—176
Callipers, Jenny, 193—194
Compressed Air, 226—232
Circular Saw, 237—239

D
Drummond Lathe, 19—20
Drummond Toolpost, 20
Drilling Machine, The, 32—40
Drilling Machine, 'Champion', 32—35
Drilling Machine 'Model Engineer', 32—36
Drilling Machine 'Cowell', 32—33, 36
Drilling Machine 'Pacera', 34
Drilling Machine, Driving the, 34—36
Drilling Machine 'Chucks', 36—38
Drilling Machine, Speed Changing Arrangements, 35—36
Drilling Machine, High Speed Attachments, 38—39
Drilling Machine, Testing the, 39—40
Dividing Head, 56
Drilling Machine Setting Ring, 143—144
Drives, V-belt, 43—44
Depth Stops, 144—145
Drill Grinding, 145—149
Drilling from the Tailstock, 84—86
Drilling Deep Holes, 86
D-Bit, 86
Dividing in the Lathe, 121—128
Dividing, 129—139
Dividing Compound, 123—124
Dividing with simple attachments, 126—128
Dividing attachment for the Headstock, 129—133
Dividing Attachment for the saddle, 133—134

Dividing, Linear, 136—138
Drills and Drilling, 140—153
Drill Stands, 141—142
Drill Chucks, 142
Drill Point, 146—149
Drill Speeds, 149
Drill Lubricants, 150
Drill, Pin, 157
Die Holder colleted, 161
Drilling Machine, Tapping in the, 168—169
Dividers, 193
Dial Indicator, 203—208
Dial Indicator Magnetic Base for, 203
Dial Indicator, Internal Attachment, 204—205

E

Electric Power Supply, 9
Electric Cables, 9—10
Electric Plugs, 9
Electric Motors, Cooling of, 14
Electric Drill, Low Voltage, 12
End Mills, Using, 114
Equipment Measuring, 180—185
Errors in Marking Out, 187—188
'Eclipse' Magnetic Base, 203
Electric Muffle, 223—225
Enamelling, Stove, 232—233

F

Furniture, Workshop, 7
Front Tool, 70
Fasteners, Belt, 41—42, 45—46
Flycutting, 110—112
Feed Screws, Independent, 214—215

G

Grinding Rest, Angular, 75—76
Grinding Wheels, 76—78
Grinding Wheels, Tuning, 77—78
Grinder, Bench, 16—18
Guard, Leadscrew, 29
Grinding Drills, 146—149
Gearing, Change Wheel, 173—174
Gear Train, Proving the, 173—174
Gauges, Small Hole, 181—182
Gauges, Depth, 185
Gauges, Taper, 185
Gauges, Slide, 183—184
Gauges, Surface, 191—193
Gauges, Scribing, 193

H

Heating the Workshop, 7—8
Headstock Mandrel, 27
Hand Tools, 74
Head, Dividing, 56
Hand Knurl Wheel Holder, 79
Holes, Drilling Deep, 86
Headstock Dividing Attachment, 129—133
Headstock Spindle, High Speed, 215—216

Hard Soldering, 219—220
Hard Soldering Equipment for, 220—221
Hardening, Case, 223—225
Hacksaw Machine, 235—237

I

Indicator, Thread, 172—173
Independent Feed Screws, 213—214

J

'Jacobs' Chuck, 37
Jacks, Screw, 190—191
Jenny Callipers, 194—195
Jenny Callipers, use of, 196—198
Jigsaw, 237—239

K

Knife Tool, 71
Keyless Chucks, 61—62
Knurling, 79—83
Knurl Wheel Holder, Hand, 79
Knurling Tool, Single Wheel, 80
Knurling Straddle, 80
Knurling Operation, Starting the, 81—82
Keyways, Cutting Internal, 95
Keyways, Marking out internal, 200—201
Keyways, Angular Location of, 200

L

Lighting the Workshop, 8—9
Low-Voltage Power Supply, 11—16
Low Voltage Electric Drill, 12
Lathes, 19—31
Leadscrew, 24—29
Lathe, Levelling the, 27—28
Lathe, caring for the, 28—31
Lathe, bearings, 29
Lathe, bed, 29
Leadscrew guard, 29
Lathe Tools, 70—78
Lathe, Drummond, 19—20
Lathe Operations, 84—95
Lapping, 102—106
Lapping of shafts, 102—103
Lapping, Internal, 103—105
Lapping compounds, 105
Lap, Boyar Schulze, 104
Lapping, Protection of Machine, 105—106
Lathe, Milling in the, 110—120
Lubrication, Oil Mist, 211
Lathe Overhead Drives, 212—216
Lathe Mandrel Change Wheel Mounting, 125—126

M

Motor generator, 12
Motors, cooling of, 14
Motors, internal connections, 14—15
Morse Taper, 26
Milling Machine, the, 52—56

INDEX

Machine Vices, 53
Machine Vices, errors in, 53—54
Machine Vices, testing, 54
Milling Machine, Tom Senior, 54—55
Milling Machine, Vertical, 55
Mandrels, 67—69
Mandrels, Plain, 66
Mandrels, Expanding, 66—67
Mandrels, Stub, 67
Mandrels, Le Count, 66—67
Mandrels, Hollow, 68—69
Mandrel, Headstock, 27
Milling in the Lathe, 110—120
Milling Attachments, 118—119
Mandrel Handle, 168
Micrometer, 180
Micrometer Stands, 181
Micrometer, Depth Gauge, 185
Measuring Equipment, 180—186
Marking Out, 186—201
Marking Out, Errors in, 187—188
Marking Out Internal Keyways, 199—200
Mist Lubrication, Oil, 211
Muffle, Electric, 223—224
Machine Tools, Additional, 235—240
Machine, Hacksaw, 235—237
Marking-out Tables, 188
Machine Reamers, 250

N
—

O
Oil Mist Lubrication, 211
Overhead Drives, Lathe, 212—216
Overhead Drives for Drummond Lathe, 212
Overhead Drives for Myford Lathe, 213

P
Power Supply, 9—13
Parting Tools, 72—73
Plugs, Electric, 9
Power Supply, Low Voltage, 11—16
Pin Drill, 157
Plates, Surface, 188
Plates, Angle, 188—189
Packing Strips, 189
Punches, Centre, 194—195
Painting, Spray, 230—231
Paint Tables, Rotary, 234

Q
—

R
Raw Material Storage, 6
Rectifiers, 11
Reversing Switch, 15
Rack, 26
Rotary Table, 56
Racks, Cutting, 136—138
Rotary Table, 138—139

Rules, Steel, 185
Raising Blocks, 189
Rotary Paint Tables, 234
Reamers, 249—252
Reamers, Adjustable, 249—250
Reamers, Machine, 250
Reamers, Taper, 250
Reamers Taper, Pin, 250—252

S
Storage of Tools, 6
Storage of Raw Materials, 6
Storage of Nuts and Bolts, 6—7
Spot drilling attachment, 12
Switches, reversing, 15
Switches, Heavy Current, 15
Switches, Foot, 16
Steadies, 21—22
Shaping Machine, the, 47—49
Shaping Machine, The Drummond Hand, 47
Shaping Machine, The Cowell Hand, 47—48
Shaping Machine, The Acorn Tools Power, 47—49
Shaping Machine, Operating the, 49—50
Shaping Machine Tools, 49
Shaping Machine Toolholders, 50
Screw Cutting Tools, 72
Self-centring chuck, 58—59
Small Tools, Boring Bar for, 89
Saws, Circular, 114—116
Slides, Vertical, 115—116
Spotfacing, 158
Screw Threads, cutting, 159—169
Squares, 193
Squares, Combination, 193
Suds Equipment, 209—211
Saddle, Boring work on the, 86—88
Small Tools, Boring Bar for, 89
Setting Ring, Drilling Machine, 143—144
Stops, Depth, 144—145
Spotfacing, 153
Screw Cutting Tools, 170—172
Screw Cutting, Practical, 176—179
Small Hole Gauges, 181—182
Steel Rules, 185
Slide Gauge, 183—184
Surface Plates, 188
Screw Jacks, 180—181
Scribers, 194
Surface Gauge, use of, 198—199
Soldering and Brazing, 217—221
Solders and Fluxes, 217
Self-Blowing Blowpipe, 217—218
Soldering, Hard, 219—220
Spray Painting, 230—231
Stove Enamelling, 232—233

T
Tool Storage, 6
Toolposts, 21
Tool Turrett, 21

Tumbler Gear, 24—25
Top Slide, 26
Taper Gauge, 50
Tool Holders, 74
Tool Grinding, 74—78
Tools, Boring, 71—72
Toolposts, Back, 73—74
Trays, drip, 30—31
Toolpost, Drummond, 20
Testing the Drilling Machine, 39—40
Tools, Hand, 74
Tool, Front, 72
Tool, Parting, 72—73
Tool, Screw Cutting, 72
Taper Turning, 96—101
Taper Turning Attachments, 96—97
Tailstock, Adjustable Centre for, 97—99
Tapers, setting with Dial Indicator, 100
Tapers, Calculating, 100
Tapers, Checking, 101
Toolmakers, Buttons, 107—109
Toolmakers Buttons, Setting, 107—109
Tap Wrenches, 162
Threading in the lathe, 162—167
Tapping in the Drilling Machine, 168—169
Tools, Screw Cutting, 170—172
Thread Indicator, 172—173
Taper Gauges, 185
Tables, Marking Out, 188
Tailstock Drilling Spindle, High Speed, 215
Toolpost, Back, 241—248
Toolpost, Back, Tools for, 246—247

U

V

Vice mounting, 6
V-belt Drives, 43—45
Vertical Slides, 116—117
V-Blocks, 190

W

Wheels, Grinding, 86—88
Workshop, Heating the, 7—8
Workshop, Lighting the, 8—9
Workshop Furniture, 7
Wrenches, Tap, 162
'Wobbler' 206—207